# CHEMICAL CONTAMINATION
AND ITS VICTIMS

**Recent Titles from Quorum Books**

The Marketer's Guide to Selling Products Abroad
*Robert E. Weber*

Corporate Social Responsibility: Guidelines for Top Management
*Jerry W. Anderson, Jr.*

Product Life Cycles and Product Management
*Sak Onkvisit and John J. Shaw*

Human Resource Management in the Health Care Sector: A Guide for Administrators and Professionals
*Amarjit S. Sethi and Randall S. Schuler, editors*

Export Strategy
*Subhash C. Jain*

The Protectionist Threat to Corporate America: The U.S. Trade Deficit and Management Responses
*Louis E.V. Nevaer and Steven A. Deck*

Environmentally Induced Cancer and the Law: Risks, Regulation, and Victim Compensation
*Frank B. Cross*

Legal and Economic Regulation in Marketing: A Practitioner's Guide
*Ray O. Werner*

Smoking and the Workplace: Issues and Answers for Human Resources Professionals
*William M. Timmins and Clark Brighton Timmins*

Private Pensions and Employee Mobility: A Comprehensive Approach to Pension Policy
*Izzet Sahin*

Group Decision Support Systems for Effective Decision Making: A Guide for MIS Practitioners and End Users
*Robert J. Thierauf*

Deregulation and Competition in the Insurance Industry
*Banks McDowell*

# CHEMICAL CONTAMINATION AND ITS VICTIMS

## Medical Remedies, Legal Redress, and Public Policy

EDITED BY
David W. Schnare and Martin T. Katzman

**Q**

QUORUM BOOKS
New York • Westport, Connecticut • London

363.179
C517

**Library of Congress Cataloging-in-Publication Data**

Chemical contamination and its victims : medical remedies, legal
  redress, and public policy / edited by David W. Schnare and Martin
  T. Katzman.
      p.   cm.
    Includes bibliographies and index.
    ISBN 0–89930–428–1 (lib. bdg. : alk. paper)
    1. Toxicology.  2. Hazardous waste—Environmental aspects—United
States.  3. Liability for hazardous substances pollution damages—
United States.  I. Schnare, David W.  II. Katzman, Martin T.
RA1216.C47  1989
363.1'79—dc19              88–37436

British Library Cataloguing in Publication Data is available.

Copyright © 1989 by David W. Schnare and Martin T. Katzman

All rights reserved. No portion of this book may be
reproduced, by any process or technique, without the
express written consent of the publisher.

Library of Congress Catalog Card Number: 88–37436
ISBN: 0–89930–428–1

First published in 1989 by Quorum Books

Greenwood Press, Inc.
88 Post Road West, Westport, Connecticut 06881

Printed in the United States of America

The paper used in this book complies with the
Permanent Paper Standard issued by the National
Information Standards Organization (Z39.48-1984).

10  9  8  7  6  5  4  3  2  1

# Contents

|    | ILLUSTRATIONS | vii |
|----|---|---|
|    | PREFACE | ix |
| 1. | The Unmapped Labyrinth: An Introduction<br>*David W. Schnare and Martin T. Katzman* | 1 |
| 2. | The Pale Light of Science: Examining the Toxicology of Low-level Exposures<br>*David W. Schnare* | 13 |
| 3. | The Limitations of Summary Risk Management Data<br>*C. Richard Cothern and David W. Schnare* | 31 |
| 4. | First Do No Harm: Diagnosis and Treatment of the Chemically Exposed<br>*David E. Root and David W. Schnare* | 53 |
| 5. | The Toxic Tort Is Ill: Deficiencies in the Plaintiff's Case and How to Prove Them<br>*Barbara P. Billauer, Avraham C. Moskowitz, and Karen L. I. Gallinari* | 65 |

## CONTENTS

6. Trans-Science in Toxic Torts — 85
   *Wendy E. Wagner*

7. Adapting Workers' Compensation to the Special Problems of Occupational Disease — 105
   *Laurence Locke*

8. Pollution Liability Insurance as a Mechanism for Managing Chemical Risks — 125
   *Martin T. Katzman*

9. The Common Law under Challenge: Politics, Science, and Toxic Torts — 145
   *James M. Strock*

10. Conclusion — 165
    *Martin T. Katzman and David W. Schnare*

    INDEX — 175

    ABOUT THE CONTRIBUTORS — 179

# Illustrations

## FIGURES

| | | |
|---|---|---|
| 2.1 | The Disease Hierarchy | 14 |
| 3.1 | Contributions to Uncertainty | 32 |
| 3.2 | The Benefits and Costs of Alternative Drinking Water Standards for TCE | 34 |
| 6.1 | The Trans-Science Time Lag | 88 |

## TABLES

| | | |
|---|---|---|
| 2.1 | The Progression of Disease Due to PCBs and PBBs | 17 |
| 2.2 | Symptoms Reported as Associated with Environmental Chemical Exposures | 22 |
| 3.1 | Uncertainty Contributions to Risk Assessments | 33 |
| 3.2 | Data Limitations and Corresponding Science Policy Options | 36 |
| 4.1 | Symptom Prevalence of Chemically Exposed and Unexposed Populations | 61 |
| 4.2 | Percent Reductions in Adipose Tissue Concentrations | 62 |
| 8.1 | Strains on the Tort and Insurance Systems from Chemical Catastrophes | 133 |
| 8.2 | Innovations Induced by the New Toxic Torts | 137 |
| 10.1 | Societal Risk Management: Regimes and Objectives | 170 |

# Preface

We do not live in a risk-free world, and we never will. As we adapt the resources of this planet to our needs, we will "bump elbows" with others, and some of those "bumps" will involve chemical exposure—intended or otherwise. This book discusses recovery from some of those bumps, and how to recognize that, in some cases, we will just have to take our bumps as they come.

Like most books about the human condition, this one describes a dynamic society. Some of the facts herein will change over the next several years. New action by Congress or the courts may make some points moot. Overall, however, the pace of action that leads to balancing the ledgers of chemical victims is a slow one. What you will find here is that toxic torts are but one aspect of a long-standing legal response to chemical exposure, and that the legal aspect is but one thread in a fabric that reflects science, medicine, insurance, and the government as well.

We have tried to weave these threads into whole cloth. The result honestly reflects the American society—it is a bit disconnected, with some threads too short, others too long, and many of clashing colors. Nonetheless, it's honest cloth, and we hope you will pull some of these threads to see where they lead, and how they connect to your own.

# CHEMICAL CONTAMINATION
AND ITS VICTIMS

# 1
# The Unmapped Labyrinth: An Introduction

## DAVID W. SCHNARE AND MARTIN T. KATZMAN

Mary Peters sits down to watch the six o'clock news. In her eighth month of pregnancy, she is as comfortable as can be expected. The pregnancy has gone well, her husband has a good job, and she has a positive outlook toward the future. Ten minutes later she is horror struck. She has learned that the grain products she has used for years contain a potent mutagen. In a second story she hears that the field behind her home was formerly a chemical waste dump and contains cancer causing chemicals. The news reporter speculates that the chemicals came from a nearby manufacturing plant—the one where Mary's husband is employed. In less than ten minutes, Mary Peters has been given just enough information to fear for the health of her unborn child, her husband, and herself. She has been given no information on her actual risk, or on how to respond to these chemical exposures. Mary Peters has been plunked down into a scientific, medical, legal, and financial labyrinth without a map. And she is not alone.

The purpose of this book is to create a map, to trace the outlines of this labyrinth, and wherever possible to show the exits from the maze. This effort grew out of a symposium sponsored by the American Association for the Advancement of Science. The purpose of the symposium was to bring together experts from all relevant disciplines to discuss five basic questions, the same questions our fictitious Mary Peters could be expected to ask:

1. What harm will befall an exposed individual?
2. What legal and medical remedies are available?
3. Who should receive remedy and compensation?

4. Who should pay?
5. How should compensation be sought?

In a simpler, nonchemical world these questions are not hard to answer. If Mary fell from a ladder while at work and broke her leg, the answers would be: (1) six weeks in a cast; (2) standard medical treatment with time off for recovery, partial payment for lost wages, and an opportunity to sue the ladder manufacturer for pain and suffering; (3) Mary gets compensation, but her workmate who used the ladder but did not fall gets no compensation; (4) the employer's insurer pays medical costs and lost wages; and (5) these payments are made through the Workers' Compensation program.

If we return to the chemical exposure problems Mary learned about on the six o'clock news, we find that these questions are not so easy to answer. (1) The harm to Mary, her baby, or her husband may not arise for over twenty years, and may be anything from birth defects or cancer to subtle neurological problems like memory loss or muscle weakness. No harm might arise at all. (2) Her family may seek treatment to reduce their chemical body burdens. However, she may not even be able to confirm their exposure. Without data on exposure, she will find no legal remedies. As such, there is no certainty that she will receive any compensation. (3) There are no set rules on who gets compensation. Mary might not while her neighbor may, or vice versa. (4) If anyone pays, it might be a chemical company, it might be a waste facility operator, it might be their respective insurance companies, in some states it might even be the government. (5) As for how best to seek compensation, the current answer seems to be "all of the above," including workers' compensation, toxic torts, out-of-court settlements, health insurance, and victims' compensation funds.

Mary isn't the only person who must confront this labyrinth. Her physician, lawyer, and insurance company, her husband's boss, union officials, workers' compensation managers and case workers, the county, state, and federal health officials, and even Mary's political representatives travel within this maze. In this opening chapter we will trace the outlines of the labyrinth, indicating the size of the chemical victim problem, its complexity and interdisciplinary nature, the response of business, government, and the courts, and the philosophies that can be brought to bear. In the succeeding chapters the main paths through the maze are laid out by representatives of the basic disciplines. At the end of this book we will reflect on where those paths connect, and how those connections can lead to real solutions. Keep in mind, however, that there are places within the maze from which there is no exit. We will close with suggestions on how a combination of disciplines might craft an opening from these unhappy corners.

## CIRCUMSCRIBING THE LABYRINTH

### Chemical Exposure

Over 7.5 million people live in the state of Michigan. In 1973 more than 97 percent of them were contaminated by a chemical fire retardant that entered the

## INTRODUCTION

food chain through cattle feed.[1] One small company was the sole responsible party for this mass poisoning. The facts of this case make it plain—we live in a chemical world, and no amount of preventive environmental action will eliminate the possibility of exposure to hazardous chemicals. Unfortunately, these chemical events are common.

In fact, the magnitude of actual chemical exposure is enormous. The number of people chemically exposed in Michigan is small compared to the documented human exposure to the pesticide DDT. Only one in a hundred U.S. citizens do not carry DDT in their bodies. This includes children who were not even born ten years ago when the chemical was banned for use in this country.[2] Summed up, that is about 233 million people exposed to DDT.

Unhappily, DDT is not the only chemical found in the bodies of the citizenry. Over 400 have been identified in human tissues, 48 in adipose tissue (fat), 40 in mother's milk, 73 in the liver, and over 250 in the blood.[3] It is certain that this list is far from complete. By 1965 over 4 million distinct chemical compounds had been reported in the literature, and about 6,000 new compounds are added to the list each week. Of these, over 65,000 are currently in commercial production.[4]

It is likely that every single individual in the United States today has been exposed to a chemical that poses some predictable risk to well-being. Further, there is a growing belief that exposure to these chemicals in consumer products, the workplace, and the environment is a significant cause of disease in our society, with cardiovascular disease and cancer the leading environmental illnesses.[5] Were it possible to identify what chemicals cause this harm, it is likely that every single individual in the United States could find someone to sue over these exposures.

It seems reasonable to ask, why aren't we all sick or dying. And why aren't we all in court? In fact, there has been a precipitous increase in personal injury suits against the manufacturers of chemicals.[6] Attempts to compensate victims of chemical exposures under the traditional tort system have been criticized by some as offering too little, too late. By others the system is seen as overcompensating dubious claimants, unfairly rewriting the rules under which business decisions are made, and destroying the insurability of catastrophic liabilities. Dissatisfaction with a struggling compensation system is inexorably drawing government into the labyrinth.

### The Missing Links

Compensation systems must judge the merits of particular claims on the basis of available scientific knowledge. Unhappily, adequate data is usually not available, complaint of harm alone being insufficient. After all, a claim of exposure is not identical to a proven exposure, and exposure to hazardous chemicals is not identical to injury. One seeks a causal relationship between a specific chemical exposure and a specific illness in a specific individual.

Ideally, one wants evidence of (a) exposure to a chemical (b) at levels that have been shown to (c) cause a human disease state (d) that is observable in the exposed individual. To get this information, one looks to toxicology, the science that studies the relationships between specific chemical doses and specific maladies.

While toxicology is a venerable science, its findings have hardly kept pace with chemical development. Of the 65,000 chemicals on the American market, the National Academy of Sciences indicates that there is "adequate" toxicological data on only one in five.[7] Even with health data on 14,000 chemicals, there are fewer than 500 for which there is a method available to measure the chemicals in human tissues. Routine analysis for diagnostic purposes is available on less than 100 chemicals.

The compensation systems turn to the toxicologists and environmental scientists for verification of chemically caused illness. It should come as no surprise to find the scientific community wandering the labyrinth with everyone else.[8]

## Institutional Disarray

The missing links in the chemical victim's chain of events causes disarray for all of the crucial actors in the drama of chemical victimization. These include the victim, the medical practitioner, the toxicologist, the lawyer, the businessman, the insurer, and the regulator. All of them must make decisions and judgments in the face of uncertainty.

Consider how confusing life can be for those who suffer physically and emotionally from apparent chemical exposure. First, the victim exposed to chemicals has difficulty getting an assessment of the risks she faces. While there are poison control centers, which give information on acute, high-level exposures, there are few local organizations that have data on low-level or chronic exposures. Furthermore, the scientific and health community is still searching for an effective means to communicate risk principles outside its immediate circle.

The medical community has difficulty recognizing the initial symptoms resulting from chemical exposure, which are usually general, nonspecific, and subtle. Physicians are currently split into three camps on environmentally engendered disease. The occupational medicine specialty has taken the lead in documenting both the subtle and obvious adverse health effects of chemicals. These physicians, however, interact with the work force, not the general public. Moreover, many of them are legally viewed as servants of the employer, who screen for potentially hypersensitive claimants, not as participants in a legally recognized doctor-patient relationship.[9]

Allergists, a second board-certified specialty, have been the traditional source of advice on nonspecific symptoms that are considered physical in origin. Allergists routinely dispatch those chemically exposed who do not respond to allergy-based treatment to the psychiatric community. Psychiatrists generally return these patients to the referring physicians, recognizing that the symptoms are not treatable with psychiatric techniques.

Having forfeited their leadership in diagnosing and treating environmentally caused diseases, the allergists are finding, to their consternation, that the patient community is seeking help from a medical subspecialty that has developed over the last decade—environmental medicine. The innate nonspecificity of chemically related diseases has forced the environmental medicine community to be highly innovative and willing to examine areas of medicine with which the other subspecialties are unfamiliar and notably uncomfortable.

The disarray in the medical community means that the chemically exposed individual has no authoritative source of information on the cause of illness, appropriate treatment, or future risk of more severe disease. And even if the individual contracts a classical chemical disease, arguing physicians will ensure that the victim has no certainty of receiving just compensation.

In addition, ignorance of specific toxicological relationships makes it impossible for compensation systems to determine whether a particular individual was harmed by a particular low-level exposure incident. But such particulars are the heart of compensation systems. A plaintiff's lawyer must prove that his client was exposed to a particular chemical and that this particular exposure, not others, caused harm.

Although the workers' compensation system has a well-established procedure for dealing with acute injuries, it is not a significant avenue of compensation for occupational disease. Unlike injury claims, most disease claims are usually litigated, and only 5 percent of those filing claims actually receive benefits. One prominent labor attorney labels this system "a crap shoot. Someone might get $500,000 and someone else might get zero, and even the trial lawyers will tell you that the person getting zero may have been just as sick."[10] Arm in arm with his medical colleagues, the compensation lawyer also travels within the labyrinth.

Not surprisingly, many victims have turned to the courts for relief. With little scientific data and a disorganized medical community, there is no reason to expect the courts to produce a stable compensation policy that results in uniform and predictable awards. In fact, judges have thrown out cases on procedural rules, even where defense counsel admit that plaintiffs have suffered real harm from chemicals. On the other hand, judges have awarded large sums to groups who have not suffered any recognizable disease, on the presumption that they face potential future risks and hence deserve compensation.

Despite the large variance in outcome, recoveries awarded to chemical victims in the aggregate can be enormous. Claims paid in 1985 for chemical exposures occurring in the United States have been estimated at over $2 billion. Worldwide claims are expected to reach $30 billion by the year 2000.[11] The bulk of these awards resulted from asbestos litigation. While most plaintiffs win, awards are relatively modest, about $55,000 per victim. Because of the volume of litigation, awards have totaled over $1 billion.[12] Vietnam veterans exposed to the defoliant

Agent Orange were awarded $180 million. Households located near the Love Canal waste site were awarded $20 million. A single producer of DDT was ordered to pay $24 million to plaintiffs. Manufacturers of dioxin lost a case asking for $2.6 million.[13]

At some point, it becomes reasonable to ask whether the chemical industry can even afford to defend itself, much less stay in business. As early as 1982, asbestos manufacturer Johns Manville filed for Chapter 11 bankruptcy in the face of 16,500 individual suits filed against it. Further, the unpredictability of the compensation system obliterates rational liability planning within the chemical industry. Chemical manufacturers, for example, have to anticipate that in the distant future they may be held retroactively liable for merely making plaintiffs fearful (*Ayers v. Jackson Township*) or for injuries that showed no relationship to any chemical (the Woburn litigation). Under the tort rule of joint and several liability, chemical companies have been held liable for damages caused by defunct firms that produced chemicals similar to their own. Welcome the courts and the chemical industry to the compensation labyrinth.

Last, but certainly not least, the unpredictability of the compensation system has caused great stress on insurers. The liability insurance mechanism works best when accidents occur within a well-defined time period, when the loss is demonstrable and calculable, and when damage is clearly related to the behavior of the insured.

Chemical technologies create hazards that are quite different from "mechanical" technologies. Mechanical accidents are sudden. While chemical accidents may be sudden, most often they reflect long-term exposures. Mechanical accidents result in injuries that are immediately manifest, while chemical injuries more usually result in latent or subtle diseases. Mechanical technologies are visible and commonplace, allowing relatively exact calculation of their inherent risks. Chemical technologies pose "invisible" risks that usually go undetected and result in rare diseases which tend to affect only a few people. However, they also arouse dread, and courts increasingly compensate for fear of illness arising from chemical exposure.

These characteristics make it difficult for an insurer to calculate the risks inherent in a chemical environment, and we can document the transition of the insurance industry from order to chaos. In 1984 insurance premiums for chemical-related enterprises rose between 100 and 200 percent and provided only half the coverage. By the end of that year, industry spokesmen announced that higher premiums and tighter liability limits would not solve the problem. In 1985 premiums were not being renewed and new policies were not being offered. While accelerating dramatically, this is not a new trend. One asbestos manufacturer has obtained liability insurance from ten different companies over forty years. Today, however, asbestos liability, unlike asbestos itself, is too hot to handle. After 1986 most businesses generating chemical wastes were forced to establish financial guarantees in lieu of insurance. Others were being forced out of the marketplace.

None of these responses offers an acceptable solution to the chemical liability problem. Leaving the market deprives consumers of the benefits of useful chemicals. Going without insurance or hiding behind the bankruptcy shield deprives potential victims of any realistic avenue for seeking compensation.

Both the chemical industry and its insurers recognize that merely raising premiums will not solve the insurance availability problem. Their frustration with the unpredictability of the compensation environment has resulted in proposals for statutory solutions, including tort reform. Chemical manufacturers favor regulations that will state unequivocally, once and for all, what acceptable practices will free them from future litigation.[14] Ignorance of these general relationships poses a serious barrier to the regulator who attempts to establish acceptable standards of exposure.

The labyrinth, we find, encloses all the players, each seeking a path leading from confusion to rationality.

## Utopia

Somewhere outside this labyrinth is a place we could call utopia. In this utopia, there are answers to all questions, treatments for every disease, fair compensation for every harm suffered; in essence, truth and justice. This utopia is not just a comic book concept. Belief in utopian solutions is part of the basic fabric of the Western world today.

The American form of utopianism assumes that science and technology can provide solutions to all problems. Physicians look to scientists to document diseases, prognoses, and treatments. Industry, government, and the general public expect scientists to create new chemicals, measure them in every conceivable medium, and explain their risk attributes. Lawyers and the courts want all this and more. Having gotten data on what risk a chemical may pose, lawyers want data on the value of avoiding that risk. Social scientists are expected to put dollar values on physical damage, pain and suffering, and even the emotional trauma of worrying about potential but as yet unrealized future harm. Even the scientific community bases its actions on this utopianism. Scientists presume that, given enough time and money, a good scientist can solve any problem.

It doesn't matter whether scientific utopianism is a valid philosophy on which to base a civilization. The fact we must face today is that utopianism is a philosophy, not a reality. Plain and simple, science does not now have the answers we would like to have in order to compensate all victims of chemical exposure. Further, science is not going to produce them any time soon.

The chemical victim labyrinth is found not in utopia, but in the realm of trans-science, to use Alvin Weinberg's phrase.[15] It is a nonutopian world that must struggle daily, if only not to lose a step. With every passing day that the various labyrinth travelers unsuccessfully labor toward rationality, the cry for government action increases in volume.

## The Role of Government

The idealist would suggest that when major elements of a society are in disarray, or no longer cooperate, it is appropriate for government to intercede. The realist would suggest that when the public voices a widely agreed-upon consensus, as documented in surveys, lobbying, or letter-writing, it is inevitable the government will intercede. When considering victims' compensation for chemical exposure, both prerequisites to government action exist. This does not guarantee that a federal or governmental highway project can or will blaze a four-lane path out of our labyrinth. However, the government has already sent in surveying teams to chart the territory.

The government is not new to victims' compensation. There are already programs to compensate victims of black lung disease and exposure to Agent Orange. In California there is a program in place to compensate people exposed to hazardous chemicals. Unfortunately, these programs are not considered ideal models.

The black lung program was expected to cost about $200 million a year. It now costs ten times that amount, $2 billion a year, even though there has been a substantial reduction in eligibility for the program. The black lung sufferers are thought to be a small group compared with chemical victims, and there is very real fear that a broader compensation program could bankrupt the government.

It is not only the cost that is a barrier to a governmental program. The same barriers to compensation that plague the courts also plague a general compensation program. The California compensation program has failed to address any low-level chemical exposure incidents and has been remarkably unable to provide any form of routine compensation for hazardous chemical exposure. The basic barrier remains the lack of an established procedure to correlate exposure to health effects, and the subsequent lack of compensation, whether for treatment or pain and suffering.

Government, unlike science, medicine, or tort law, is not especially constrained by the lack of facts on chemical-based disease. The government can be arbitrary, as long as it is not capricious as well. For example, Congress authorized a compensation program for veterans exposed to Agent Orange, despite the clear lack of corroborative health effects data. In another case, Congress limited compensation liability for accidents at nuclear power plants. In the former case, government responded to the expressed fears of the chemically exposed, in the latter the response was to the fears of the potential contaminator.

However, for each scientific barrier the legislative or administrative branches of government breach, there is likely to be an otherwise impossible political or administrative barrier that might be unbreachable. The current nonscientific barrier to legislative action, one which is certain to be around for several years, is an unbalanced budget. We suggest, therefore, that while government is certain

# INTRODUCTION

to be with all the other players in the labyrinth, it has its own navigation problems, and does not represent the inevitable path out.

## INSIDE THE MAZE

The chapters that follow provide a ground view of the compensation labyrinth. The authors are guides who know the territory well. They draw expertise from different disciplines and perspectives in the labyrinth. In a few cases, they know how to get out on their own. Taken as a whole, we hope their insights will suggest new approaches that open the territory significantly.

In Chapter 2, David Schnare dispels any of the reader's residual scientific utopianism. Examining the data linking chemical exposure to health effects, he indicates what is known, what is not, and under what conditions new knowledge is likely to emerge. All is not gloom and doom, however, as there is a lot of useful data at hand. He shows how such data can be used, misused, and inappropriately abandoned.

Whatever its scientific status, the limited knowledge at hand will certainly be used by protagonists in the victim compensation controversy. In Chapter 3, Richard Cothern and David Schnare focus on how fragmentary knowledge is used by regulators to make decisions about acceptable risk. They show how it is possible to skirt data gaps, fill them in with assumptions, and in some cases, ignore them when reaching conclusions about health effects.

While toxicologists are becoming the darlings of the legal community, they remain a few steps removed from the public. Both the general public and the insurance communities look to the physician for diagnosis and treatment of the consequences of chemical exposure. When the toxicologist is unable to assess the potential risks of a chemical exposure, the physician is on his own. While inability to prove causation scientifically may now bar an individual from recovering damages in court, physicians may still have enough information to diagnose and treat the victim successfully. In discussing the approaches that have been developed for diagnosis and treatment, David Root presents the state of the art of occupational medicine. His discussion in Chapter 4 suggests that there may be treatment-related limits to compensation for noncatastrophic, preventable, or reducible future risks. Here he discusses the techniques for reducing chemical body burdens, thereby reducing some chemical disease symptoms, as well as future risk and the associated fear of harm.

In many cases, the first trip taken after chemical exposure is not to the doctor's office, but to the lawyer's. Alleged victims of chemical exposure have sought common law remedies, with mixed success. Proving a causal link between low-level exposures to chemical hazards and disease is a major barrier to compensation. In relating chemical exposure to injury, the standards of scientific evidence employed by the courts differ dramatically from those used by regulators in deciding acceptable risk. The value of limited toxicological knowledge is viewed differently by the defendant's and plaintiff's bar.

Barbara Billauer, Avraham Moskowitz, and Karen Gallinari examine the courtroom interpretation of toxicological data from the defense perspective. In Chapter 5, they argue that the tort system has gone beyond the limits of good science. Theory should not replace fact, and the fear of harm should not be equated with harm itself. Because a reduction in the standards of scientific proof of causation results in more harm than good, bad science makes bad public policy.

The plaintiff, however, is not without a credible legal argument, despite scientific uncertainty. In Chapter 6, Wendy Wagner presents a causation standard that circumvents problems whose solutions are scientifically indeterminate by combining a qualitative showing of causation with proof that the manufacturer acted negligently in introducing an "abnormally dangerous" product. This approach would shift the burden to the defendant to prove that the product was safe, the hazards were not foreseeable, benefits outweighed potential costs at the time of marketing, or that the plaintiff was not exposed to substantial concentrations of the products.

While many chemical-injury cases must go to court for remedy, victims who can demonstrate that injuries were occupationally related can approach the workers' compensation system for redress. While the system was developed to handle sudden or acute injuries, Laurence Locke shows how the system has learned to cope with occupational disease. In Chapter 7, he argues that the workers' compensation system no longer works, and suggests reforms that address the unique problems posed by chemically caused occupational disease.

Regardless of how chemical injuries are litigated, through tort or administrative systems, recoveries are usually financed by a third party—the insurance company. Requiring chemical manufacturers and hazardous waste disposers to prove "financial responsibility" through insurance helps guarantee that the chemical industry has the funds to compensate successful plaintiffs. In Chapter 8, Martin Katzman traces the collapse of the pollution liability insurance market back to the unpredictability of legal rules, which have evolved to deal with the scientific uncertainties. The implications of the collapse of this market on victim compensation and efficient risk management are discussed.

Through the first eight chapters we provide a view of how professionals cope with scientific uncertainties in the tort, administrative, and regulatory arenas, as well as how such coping has affected those who seek and those who pay damages. In the final two chapters we look at potential reforms.

In Chapter 9, James Strock reviews the congressional perspective on victim compensation. While the major elements of potential government action reflect the barriers identified in the preceding chapters, the initiating bases for government action are raw political forces. These forces may have as significant an effect on the eventual outcome as the basic compensation issue itself.

In the concluding chapter we reexamine the labyrinth, indicating how the existing institutions use fragmentary scientific knowledge to make decisions about compensation. We weigh the extent to which a more predictable and equitable compensation system could be developed with the existing scientific knowledge against the extent to which further scientific advance is necessary.

And, what about Mary Peters, whose concerns opened this chapter? Our analysis suggests there is medical treatment she can get, some compensation she might receive, and some fear that may be allayed. Like hundreds of millions of other Americans, however, her risk of chemical disease is unknown and will probably remain so throughout her life.

## NOTES

1. M. Wolff, H. Anderson, and I. Selikoff, "Human Tissue Burdens of Halogenated Chemicals in Michigan,"*Journal of the American Medical Association* 247 (1982):2112.

2. F. W. Kutz, S. Strassman, and J. Sperling, "Survey of Selected Organochlorine Pesticides in the General Population of the United States: Fiscal Years 1970–1975," *Annals of the New York Academy of Science* 320 (1979):60–68.

3. U.S. Environmental Protection Agency, *Chemicals Identified in Human Biological Media, A Data Base*, EPA 560/13–80–036B, PB 81–161–176, (Washington, D.C.: 1980).

4. Council for Environmental Quality, *Environmental Quality* (Washington, D.C.: GPO, 1976).

5. Rene Zenter, "Hazards in the Chemical Industry," *Chemical and Engineering News* 57 (1979):25–34.

6. P. Huber, "Injury, Litigation and Liability Insurance," *Science* 2 (1987):31.

7. National Research Council—National Academy of Sciences, *Toxicity Testing: Strategies to Determine Needs and Priorities* (Washington, D.C., 1984).

8. K. Schneider, "The Data Gap," *The Amicus Journal*, Winter 1985, pp. 15–24.

9. Mark A. Rothstein, *Medical Screening of Workers* (Washington, D.C.: Bureau of National Affairs, 1984), Chapter 1.

10. R. Stanfield, "Toxic Tort System: A Costly Gamble For Victims And Manufacturers," *National Journal*, 5 October 1985, pp. 2250–54.

11. Ibid.

12. J. S. Kakalik et al., *Variation in Asbestos Litigation Compensation and Expenses* (RAND Institute for Civil Justice, 1984).

13. Stanfield, "Toxic Tort System."

14. For example, "Limit Landfill Use, CMA Spokesman Urges," *Chemecology*, February 1983, p. 3.

15. A. Weinberg, "Science and Its Limits: The Regulator's Dilemma," *Issues in Science and Technology* 1 (1985):59–72.

# 2

# The Pale Light of Science: Examining the Toxicology of Low-level Exposures

## DAVID W. SCHNARE

Toxicology, put simply, is the study of poisons. We spend hundreds of millions of dollars each year in a continuing effort to define the toxicology of chemical exposure. A single two-year animal study results in over 250,000 tissue samples and over a million data elements, all aimed at describing the poisonous effect (especially the potential for cancer) associated with exposure to a single chemical.

What is the result of all this effort? Dexter Goldman of the Environmental Protection Agency says: "There are no answers in toxicology, only opinions."[1] In other words, he feels we may pursue all this work but we cannot make definitive conclusions about the causal link between chemical exposure and disease.

Fortunately, Goldman's statement reflects only a small portion of the disgruntlement within the society occasioned by the scientific community's inability to define exactly the causes of all disease, and of course the concomitant cures for all disease.

This chapter will break the subject of chemical-related diseases into two basic types: those on which the scientific community has passed judgment and those on which they have not. Different approaches for applying the fruits of toxicology are available under each scenario.

In the former case, the physician community can take direct action, the legal community can craft a stable liability response, and the insurance community can accumulate loss experience that is useful in setting premiums. In the latter, the scientific community will push forward, but solutions to the problem must await significant new scientific consensus or must lie outside the scientific milieu.

**Figure 2.1**
**The Disease Hierarchy**

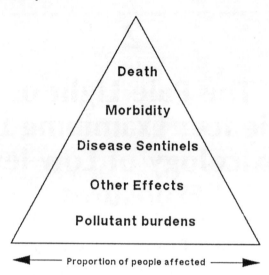

## THE DISEASE HIERARCHY

The medical definition of disease is "any deviation from or interruption of the normal structure or function of any body part, organ, or system that is manifested by a characteristic set of symptoms and signs and whose etiology, pathology, and prognosis may be known or unknown."[2] When defining diseases associated with chemical exposure, two problems arise. First, effects, which can be quite subtle, vary dependent upon dose and characteristics of the individual. Second, quite different chemicals can cause quite similar effects.

### The Hierarchy

Colucci provides us with a simple figure (Figure 2.1) that indicates the hierarchy of disease associated with chemical exposure.[3] This scale of effect severity is continuous up to the point of death. Morbidities are usually defined by clinical and subclinical signs. These may be changes at the system level, the organ level, and at the cellular level. They are generally measurable quantitatively, but may be directly observable by a clinician in a qualitative manner.

Morbidity may take the form, for example, of increased blood pressure, decreases in numbers of red blood cells, increases in white blood cells, increases in specific enzymes, decreases in the rate of nerve impulse transmission, rashes, IQ and personality trait changes, and a myriad of other clearly observable effects, all considered adverse.

Below morbidity are the subtle effects of chemical exposure. Over a decade ago Golberg defined the study of these effects as "subliminal toxicology."[4] More often than not, these effects are observable as subtle functional changes such as slowing of motor reactions, impaired regulation of appetite, reduced visual discrimination capacities, fatigue, and memory loss. As Mello points out, these "early and incipient stages of intoxications are marked by vagueness and ambiguity."[5] Further, as Weiss and Simon indicate, "these are not deficits that induce people to seek out physicians."[6]

While many of these disease sentinels are measurable by quantitative means, many are not. Consider, for example, the symptoms found to be associated with low-level exposure to PBB (polybrominated biphenyls): nausea, dizziness, depression, nervousness, paresthesia, tiredness, sleepiness, loss of balance, muscle weakness, blurred vision. While few of these can be measured, the evidence of their association with PBB exposure is unassailable, and these are certainly undesirable health effects.

At some point, the effects of chemical exposure are of uncertain importance. We know, for example, that as people age some of their neurological function is reduced. They may have some memory loss or their reaction times may deteriorate. As this is a normal phenomenon, how much more rapid must this loss of neurological function be for it to be considered an adverse effect?

Definition of "adverse effect" is a consuming issue for regulatory agencies, a point others have discussed at some length.[7] Without engaging in that discussion, there is a point at which the effects of chemical exposure are so easy to compensate for that they are considered insignificant, despite the fact that they are the earliest sentinels of disease.

## Disease and Time

A confounding element of disease is the point in time after exposure at which effects arise. Effects tend to be divided into those that arise shortly after exposure (acute effects) and those that arise long after exposure, last for long periods of time, or require long periods of exposure before they arise (chronic effects). Acute and chronic effects may or may not be significantly life-shortening. For example, lung cancer, mostly due to smoking, will often shorten life by more than a decade, while breast cancer usually shortens life by only a few months. The more acute the effect, the more likely that it can be associated with a specific chemical exposure. Chronic effects, like cancer, are difficult to associate with a specific exposure because the individual may have been exposed to a plethora of chemicals during the latency period, several of which may cause the chronic effect.

The hierarchy of disease is insensitive to the acute or chronic nature of the effect. There are cases of each all along the hierarchy of disease. However, there is a relationship between the hierarchy and the life-shortening nature of an effect.

## Progression of Disease

Underlying toxicology research is the time-tested model of biological action: the greater the chemical exposure, the greater the resultant effect. This model has been found to hold true for chemical health threats such as those we find in the environmental and occupational arenas.

The chemical exposure threat to an individual is best thought of as the dose to the sensitive organ or biological system. Therefore, the blood level of a chemical can be an important indicator of the threat of a chemical to an organ that is heavily perfused by blood. Or, if the threatened system is small or has little regenerative capacity, a dose delivered to that system (or stored in that system) can be quite small but quite significant.

The nervous systems are an example of the latter case. There are few nerve cells in the body in comparison with the number of cells in other organs or systems. Further, nerve cells do not regenerate, and neural pathways are not easily replaced. It is not unexpected, therefore, that investigators such as Spyker write: "There is growing evidence that nervous tissue, especially the brain, is more sensitive to many foreign chemical substances than has previously been suspected, and that toxic effects may be manifested as subtle disturbances of behavior long before any classical symptoms of poisoning become apparent."[8]

This was documented in the enormous PBB contamination incident in Michigan in the early 1970s, and is true for lead, mercury, acrylamide, organophosphates and a host of other chemicals.[9]

These early sentinels of chemical-related diseases can be easily ignored, especially if not looked for. Scharnweber points out that in the more common cases of chronic intoxication, all clinical tests are usually negative, despite the clear and plainly undesirable symptomatology.[10]

Table 2.1 displays the progression of diseases in humans exposed to PBB and PCB.

As expected, the first indications of chemically caused disease are neurological. As exposures increase, clear clinical effects can be found. At still higher levels the effects of systemic failure or noncompensation arise. Eventually the dose rises to a point where death is the immediate result of the exposure.

Table 2.1 displays an interesting point to note. At fairly low body burdens of levels similar to those that cause subtle neurological effects, evidence is available to document carcinogenic effects.[11]

With this progression in mind, the health-provider community would want to be able to engage in disease diagnosis at the earliest point in time. This presents a diagnostic challenge of no small proportion.

## Similarity in Disease Sentinels

If every chemical exposure had a unique biological response, it would be much easier to relate a chemical exposure to a disease state. Unfortunately, this

**Table 2.1**
**The Progression of Disease Due to PCBs and PBBs**

| Exposure | Tissue Concentration | Clinical Response | Biological Response |
|---|---|---|---|
| Ambient | 2.3 ppb blood<br>0.08 ppm fat | None Observed | Coping |
| Low-Level | 11. ppb blood<br>0.4 ppm fat | Subtle Symptoms | Many: e.g. fatigue nervousness joint pain impared memory etc. |
| Occupational | 90. ppb blood<br>25. ppm faat | Subclinical & Clinical Signs | Immune disfunction, Elevated CAE titer & SGOT-SGPT |
| Extended Occupational | 603. ppb blood<br>196 ppm fat | Overt Signs & Symptoms | Dermal abnormalities Abdominal pain Eye irritation |
| Massive | | Premature Death | Major systemic failures |
| Lifelong Ambient | 8.7 ppm fat | Premature Death | Cancer |

is not the case. A review of the toxicology of forty-six high-production chemicals with regard to fifty-three symptoms and signs demonstrates the overlap in symptomotology. This data set is given in the table in the appendix to this chapter that cross-references symptoms and chemical exposures. The chemicals are quite dissimilar in structure and chemical property. At high exposure levels they affect quite different organs and cause quite distinguishable diseases. However, at low exposure levels many produce very similar symptoms.

An extreme example of this was the case of several lead-poisoned workers who were hospitalized for abdominal distress, including colic, constipation, and upset stomach. Medical care was extensive, including abdominal surgery, with negligible results until the history of lead exposure was serendipitously revealed. Therapy for lead poisoning brought rapid relief, but was provided only after highly invasive, expensive, risky, and ineffective initial medical treatment.[12]

Many investigators acknowledge the ease of confusion of a chemical exposure with other common diseases or with different chemical exposures. However, they contend that with occupational information, the clinical picture is sufficiently characteristic to allow correct diagnosis, even without a definitive test.[13]

## Status of Research

The state of knowledge on the progression of disease, and causality of effect, is bounded by the problems discussed above. Mechanisms to improve data bases are in use, although perhaps not as fully as might be desired.

For example, there has been a rapid expansion of efforts to define behavioral toxicology effects of chemical exposures.[14] Work in this subject reaches back as far as 1955[15] and is an extremely active subject in the Soviet Union.[16] However, there is so much toxicology yet to be done that a means must be developed to generalize the existing data while adding to the data base itself.

## DATA AVAILABILITY

Having discussed the hierarchy of health effects and the difficulty of their observation at low doses, it is appropriate to ask how much data is available to the physician who is seeking to diagnose and treat an ill patient. It is fair to say that there is both a great deal of data, yet only a fraction of what is needed.

The National Academy of Sciences' recent review of toxicological data indicates that there is "adequate" data on over 14,000 chemicals.[17] That is a lot of data. However, there are over 65,000 chemicals on the American market, and this "adequate" data set covers only about one in five chemicals (22 percent).

Most of the data available is for commercial/industrial chemicals, and the highest degree of completeness of data is on drugs. What is most alarming in this data is the amount of missing or inadequate information on chemicals that have extremely high human-exposure potential—drugs, food additives, and cosmetics that are ingested, inhaled, absorbed, or injected.

All new drugs must undergo relatively intensive investigation before they can be marketed. The same is true for new food additives. However, before federal regulations were put into place requiring such information and approval, many existing chemicals could be found in the marketplace. Most of those were "grandfathered" in to allowed use, and few of them have been seriously examined for toxicological effect since.

The NAS study shows that data is best collected when the chemicals must be "approved" or "permitted." The majority of chemicals in the marketplace are commercial/industrial ones, and few of them need to be approved or permitted. In cases where a chemical becomes widely used, and human exposure is thought to be likely, the federal government will place the chemical in question into its queue for testing under the National Toxicology Program. The current list is extremely long, and it will be decades before funds permit testing of the existing list, much less new chemicals which could be added.

A second problem associated with the toxicology data gap is the form of data being developed. The health effects of low-level chemical exposure may be chronic and life-threatening in nature (cancer), but are first observed as more subtle debilitating symptoms. Current testing protocols do not address subliminal or behavioral toxic effects. Nor can they. It is very difficult to evaluate human behavioral effects in nonhumans, and existing protocols are based on use of rodents, for the most part.

A third problem comes from the complexity of chemical exposures. The health-effects literature is rich with examples of interactive effects of chemical exposures. In come cases health effects are worse than merely the sum of the effects expected from the exposures given. In other cases, and in far fewer cases, effects are less than expected. Overall, however, there is very little comprehensive data on synergism and antagonism of effects.

Consider, for example, the amount of research that would have to be done to specify all the possible effects of the ten most commonly found organic contaminants in drinking water. If a single dose level in a single animal species were to be tested, there would have to be 603 different experiments. If a full protocol of two species, two sexes, and three doses were used, $1.16 \times 10^{14}$ tests would have to be conducted. That is 100 thousand billion tests for a mere ten chemicals. At thirty rats and mice per sex, per dose, even the Pied Piper would be awfully busy.

It is not difficult to understand why short-term tests that are being done at the subcellular and cellular level are being investigated with fervor. However, again, these tests are aimed at cancer, not the debilitating subliminal toxic effects.

## STRATEGIES FOR ACTION

What contribution, then, does the toxicology community bring to the problem of low-level contamination. Is it as Goldman suggests, that toxicology can offer no solutions only opinions?

Certainly not. As Hanson and colleagues make clear, data on exposures is often enough to allow correct diagnosis of low-level chemical pathology even without definitive clinical tests.[18] The saving grace is the same phenomenon that tends to stand in the path of documenting causality—similarity in symptomotology.

Once the obvious alternative etiologies of symptoms are discarded (viral and bacteriological bases), the remaining probable cause is chemical. Such a diagnosis does not, however, justify unlimited medical or legal intervention.

The data available on chemical effects is sufficient to support additional diagnostic evaluation of exposed patients, as long as the diagnostic evaluations will not increase risk to the patient and are not themselves ridiculously expensive.

Because most of the chemicals that have a chronic effect (whether subliminal, behavioral, or significantly life-shortening) bioaccumulate, noninvasive treatments to reduce body burdens of chemicals are also justified, again as long as they do not increase risk to the patient and are not ridiculously expensive.

Fortunately there are such treatments available, as is discussed in Chapter 4, and elsewhere.[19]

Lastly, because the debilitating nature of the non-life-shortening health effects are often remediable, and do not constitute extraordinary conditions, relatively low-cost monetary settlements similar in size to the cost of noninvasive, body-burden reduction treatment are supportable by science.

There are, of course, medical and legal interventions that are not a reasonable extension of the available scientific data or research methods. Cancer is a probabilistic disease. Only about one in ten people who smoke cigarettes get lung cancer. While it is possible to predict how many people in a population will get cancer due to a specific exposure, it is not possible to predict which of the exposed people will be the unlucky ones. Although there are some clever ideas that may breast this scientific barrier, adequate arguments have yet to receive serious consideration.

As a result, invasive treatments that pose significant risks to the patient should not be encouraged. For example, the risks associated with general anesthesia are far greater than the probable life-shortening risks of low-level chemical exposure. In like measure, I don't believe there can be a justification for large monetary settlements where no significantly life-shortening or life-quality-ruining disease is evident.

In cases where cancer or other life-shortening disease is present, there is justification for seriously invasive diagnostic analysis. For example, the concentration of chemicals to which the patient was exposed, and which he might seek to implicate as part of a liability suit, can be examined in malignant and normal tissues of the patient. If the suspect chemicals, or their persistent metabolites are found, there may be grounds for a large monetary settlement or highly invasive treatment. Science can go that far in attempting to solve the low-level chemical exposure problem. If such chemicals are not found, despite the presence of the life-threatening disease, one must look outside the toxicological arena for an appropriate solution.

## NOTES

1. Goldman quoted in K. Schneider, "The Data Gap," *The Amicus Journal*, Winter 1985, pp. 15–24.
2. *Dorland's Medical Dictionary*, 23d ed. (Philadelphia: W. B. Saunders, 1982).
3. A. V. Colucci et al., "Pollutant Burdens and Biological Response," *Archives of Environmental Health* 27 (1973):151–54.
4. L. Golberg, "Safety of Environmental Chemicals—The Need and the Challenge," *Food and Cosmetics Toxicology* 10 (1972):523–29.
5. N. K. Mello, "Behavioral Toxicology: A Developing Discipline," *Federation Proceedings* 34, no. 9 (1975):1832–34.
6. B. Weiss and W. Simon, "Quantitative Perspectives on the Long-term Toxicity of Methylmercury and Similar Poisons," in *Behavioral Toxicology*, ed. B. Weiss and V. G. Laties (New York: Appleton-Century-Crofts, 1976).

7. W. Marcus and C. R. Cothern, "Defining an Adverse Health Effect: Using Lead as an Example" (Paper delivered at the 151st Annual Meeting of the American Association for the Advancement of Science, Los Angeles, California, May 29, 1985).

8. J. M. Spyker, "Assessing the Impact of Low-level Chemicals on Development: Behavioral and Latent Effects," *Federation Proceedings* 34, no. 9 (1975):835.

9. J. A. Valciukas et al., "Comparative Neurobehavioural Study of a Polybrominated Biphenyl-exposed Population in Michigan and a Non-exposed Group in Wisconsin," *Environmental Health Perspectives* 23 (1978):199-210.

10. See H. C. Scharnweber, G. N. Spears, and S. R. Cowles, "Chronic Methyl Chloride Intoxication in Six Industrial Workers," *Journal of Occupational Medicine* 16 (1974):112; and M. Wassermann et al., "Organo-chlorine Compounds in Neoplastic and Adjacent Apparently Normal Gastric Mucosa," *Bulletin of Environmental Contamination and Toxicology* 20 (1978):544-53.

11. M. Unger and J. Olsen, "Organochlorine Compounds in the Adipose Tissue of Deceased People with and without Cancer," *Environmental Research* 23 (1980):257-63.

12. J. D. Repko, and C. R. Corum, "Critical Review and Evaluation of the Neurological and Behavioral Sequelae of Inorganic Lead Absorption," *CRC Critical Reviews in Toxicology* 6, no. 2 (1979):135.

13. See H. Hanson, N. K. Weaver, and F. S. Venable, "Methyl Chloride Intoxication," *Archives of Industrial Hygiene Occupational Medicine* 8 (1953):328; R. E. Eckardt, "Industrial Intoxications Which May Simulate Ethyl Alcohol Intake," *Industrial Medicine and Surgery* 40 (1971):33; P. M. Fullerton, "Industrial Disease of the Central Nervous System," *Society of Occupational Medicine Transactions* 19 (1969):91; J. D. Repko and S. M. Lasley, "Behavioral, Neurological, and Toxic Effects of Methyl Chloride: A Review of the Literature," *CRC Critical Reviews in Toxicology* 6, no. 4 (1979):283; J. A. Valciukas, "Comparative Neurobehavioural Study," 200; and Scharnweber, "Chronic Methyl Chloride Intoxication."

14. C. L. Mitchell, "Assessment of Neurobehavioral Toxicity: Problems and Research Needs," *Trends in Pharmacological Sciences* 4 (1983):195-98.

15. R. Furchtgott, "Behavioral Effects of Ionizing Radiations," *Psychological Bulletin* 60, no. 2 (1963):157-99.

16. H. B. Elkins, "Maximum Acceptable Concentrations: A Comparison in Russia and the United States," *AMA Archives of Environmental Health* 2 (1961):45.

17. National Research Council-National Academy of Sciences, *Toxicity Testing: Strategies to Determine Needs and Priorities* (Washington, D.C., 1984).

18. H. Hanson, "Methyl Chloride Intoxication."

19. See D. W. Schnare, M. Ben, and M. G. Shields, "Body Burden Reductions of PCB's, PBB's and Chlorinated Pesticides in Human Subjects," *Ambio* 13, no. 5/6 (1984):378-80; and D. B. Katzin, "A Review of Diagnostics and Treatment of Patients Exposed to Low-level Contaminants" (Paper delivered at the 151st Annual Meeting of the American Association for the Advancement of Science, Los Angeles, California, May 30, 1985).

# APPENDIX

This appendix is a list of symptoms reported in the literature to be associated with exposure to some forty-six chemicals. This list draws on both review articles and primary research reports and covers articles published in the last ten years. Because researchers did not regularly assess all symptoms listed herein, there is no guarantee that these are the only symptoms caused by these chemicals.

This appendix consists of three parts. The symptoms used in the table are identified by major group in the first section. In the second section a table of chemicals versus symptoms (Table 2.2) is presented. The table entries are references to illustrative citations, the third section.

This appendix should only be considered illustrative of the data available in the literature. It is by no means a comprehensive tabulation. However, despite probable missing entries, there is a clear indication that a variety of chemicals will cause similar sets of symptoms.

**Table 2.2**
**Symptoms Reported as Associated with Environmental Chemical Exposures**

|  | EYES ||||||| SKIN ||||||||
|---|---|---|---|---|---|---|---|---|---|---|---|---|---|---|---|
|  | E1 | E2 | E3 | E4 | E5 | E6 |  | S1 | S2 | S3 | S4 | S5 | S6 | S7 | S8 |
| Acrylamide |  |  |  |  |  |  |  |  |  |  |  |  |  |  |  |
| Adiponitrile |  |  |  |  |  |  |  |  |  |  |  |  |  |  |  |
| Aniline |  |  |  |  |  |  |  |  |  |  |  |  |  |  |  |
| Arsenic |  |  |  |  |  |  |  |  |  |  |  |  |  |  |  |
| Arsine |  |  |  |  |  |  |  |  |  |  |  |  |  |  |  |
| Benzene |  |  |  |  |  |  |  |  |  |  |  |  |  |  |  |
| Bromophenylacetylurea |  |  |  |  |  |  |  |  |  |  |  |  |  |  |  |
| Carbon Disulfide |  |  |  | 3 |  |  |  |  |  |  |  |  |  |  |  |
| Carbon Monoxide |  |  |  |  |  |  |  |  |  |  |  |  |  |  |  |
| Carbon Tetrachloride |  |  |  |  |  |  |  |  |  |  |  |  |  |  |  |
| Chlordane |  |  |  |  |  |  |  |  |  |  |  |  |  |  |  |
| Chlorinated Hydroquinolines |  |  | 3 |  |  |  |  |  |  |  |  |  |  |  |  |
| Chloroprene |  |  |  |  |  |  |  |  |  |  |  |  |  |  |  |
| Cyanide |  |  |  |  |  |  |  |  |  |  |  |  |  |  |  |
| DDT |  |  |  |  |  |  |  |  |  |  |  |  |  |  |  |
| Dichloroethane |  |  |  |  |  |  |  |  |  |  |  |  |  |  |  |
| Dimethyl Sulphate |  |  |  |  |  |  |  |  |  |  |  |  |  |  |  |
| Dioxin |  | 5 |  |  |  |  |  |  |  |  |  |  |  |  |  |
| Ether, diethyl |  |  |  |  |  |  |  |  |  |  |  |  |  |  |  |
| Ethylene Dichloride |  |  |  |  | 5 |  |  |  |  |  |  |  |  |  |  |
| Hexachlorophene |  |  | 5 |  |  |  |  |  |  |  |  |  |  |  |  |

# SYMPTOMS ASSOCIATED WITH ENVIRONMENTAL CHEMICAL CONTAMINATION

**EYE**

E1        Eye irritation
E2        Dimness of sight (amblyopia)

**Table 2.2** Continued

|  | EYES | | | | | | SKIN | | | | | | | |
|---|---|---|---|---|---|---|---|---|---|---|---|---|---|---|
|  | E1 | E2 | E3 | E4 | E5 | E6 | S1 | S2 | S3 | S4 | S5 | S6 | S7 | S8 |
| Hexane |  |  |  |  | 5 |  |  |  | 16 | 16 | 16 | 16 |  |  |
| Hydroquinone |  |  |  |  |  |  |  |  |  |  |  |  |  |  |
| Kepone |  |  |  |  |  |  |  |  |  |  |  |  |  |  |
| Lead |  |  |  |  |  |  |  |  |  |  |  |  |  |  |
| Leptophos |  |  |  |  |  |  |  |  |  |  |  |  |  |  |
| Manganese |  |  |  |  |  |  |  |  |  |  |  |  |  |  |
| Mercury |  |  |  |  |  |  |  |  |  |  |  |  |  |  |
| Methyl Bromide |  |  |  |  |  |  |  |  |  |  |  |  |  |  |
| Methyl Chloride |  |  | 5 | 14 |  |  |  |  |  |  |  |  |  |  |
| Methyl-n-butyl Ketone |  |  |  |  |  |  |  |  |  |  |  |  |  |  |
| Methylmercury |  |  |  |  | 5 |  |  |  |  |  |  |  |  |  |
| Nitrofurans |  |  |  |  |  |  |  |  |  |  |  |  |  |  |
| Organo-phosphate Esters |  | 5 |  |  | 5 | 5 |  |  |  |  |  |  |  |  |
| Organo-phosphates |  |  |  | 11 |  |  |  |  |  |  |  |  |  |  |
| PBB | 1 |  |  | 15 |  |  | 1 | 1 | 1 | 1 | 1 | 1 | 1 | 1 |
| PCB | 13 |  |  |  |  |  | 6 | 6 | 6 | 6 |  |  |  |  |
| Phenylmercury |  | 5 |  |  |  |  |  |  |  |  |  |  |  |  |
| Polychlorinated |  |  |  |  |  |  |  |  |  |  |  |  |  |  |
| Polycyclics |  |  |  |  |  |  | 9 | 9 | 9 | 9 |  | 9 |  |  |
| Radiation, ionizing |  |  |  | 7 |  |  |  |  |  |  |  |  |  |  |
| Sulfur Dioxide |  |  |  |  |  |  |  |  |  |  |  |  |  |  |
| Tetraethyl lead |  |  |  |  |  |  |  |  |  |  |  |  |  |  |
| Tetraethylthiuram Disulphide |  |  |  |  |  |  |  |  |  |  |  |  |  |  |
| Thallium |  |  |  |  | 5 | 5 |  |  |  |  |  |  |  |  |
| Toluene |  |  |  |  |  |  |  |  |  |  |  |  |  |  |
| Trichloroethylene |  |  |  |  |  |  |  |  |  |  |  |  |  |  |

| E3 | Double vision (diplopia) |
| E4 | Blurred vision |
| E5 | Eye oscillation (nystagmus) |
| E6 | Pupil reactions |

**SKIN**

| S1 | Rash |
| S2 | Acne |
| S3 | Sun sensitivity |

**Table 2.2** Continued

|  | NEUROLOGICAL FUNCTION ||||||||||||||
|  | Sensory |||||| | Motor |||||||
|  | P1 | P2 | P3 | P4 | P5 | P6 |  | M1 | M2 | M3 | M4 | M5 | M6 | M7 | M8 |
|---|---|---|---|---|---|---|---|---|---|---|---|---|---|---|---|
| Acrylamide |  |  |  |  | 5 |  |  | 5 | 5 | 5 |  |  |  |  |  |
| Adiponitrile |  |  |  |  |  |  |  |  |  |  |  |  |  |  |  |
| Aniline |  | 5 |  |  |  |  |  |  | 5 |  |  |  |  |  |  |
| Arsenic |  | 2 | 5 |  |  |  |  |  | 3 |  |  |  |  |  |  |
| Arsine |  |  | 5 |  |  |  |  |  |  |  |  |  |  |  |  |
| Benzene |  |  |  |  |  |  |  | 5 |  |  |  |  |  |  |  |
| Bromophenylacetylurea |  |  |  |  |  |  |  |  | 3 |  | 3 |  | 3 |  |  |
| Carbon Disulfide |  |  | 3 | 10 |  |  |  |  | 10 |  |  |  |  |  |  |
| Carbon Monoxide |  |  |  |  |  |  |  |  | 5 |  |  |  |  |  |  |
| Carbon Tetrachloride |  |  |  |  |  |  |  |  |  | 5 |  |  |  |  |  |
| Chlordane |  |  |  |  |  |  |  |  |  |  |  |  |  |  |  |
| Chlorinated Hydroquinolines |  |  | 3 |  |  |  |  |  | 3 | 3 |  |  |  |  |  |
| Chloroprene |  |  |  |  |  |  |  |  |  |  |  |  |  |  |  |
| Cyanide |  | 5 |  |  |  |  |  |  | 5 |  |  |  |  |  |  |
| DDT |  |  |  |  |  |  |  |  |  | 5 |  |  |  |  |  |
| Dichloroethane |  |  |  |  |  |  |  |  | 5 | 5 |  |  |  |  |  |
| Dimethyl Sulphate |  |  |  |  | 5 |  |  |  |  |  |  |  |  |  |  |
| Dioxin |  |  |  |  |  |  |  |  |  |  |  |  |  |  |  |
| Ether, diethyl |  |  |  |  |  |  |  |  |  |  |  |  |  |  |  |
| Ethylene Dichloride |  |  |  |  |  |  |  |  |  | 5 |  |  |  |  |  |
| Hexachlorophene |  |  |  |  |  |  |  |  |  | 5 |  |  |  |  |  |

| S4 | Darkening or thickening |
|----|---|
| S5 | Discoloration or deformity of nails |
| S6 | Dryness |
| S7 | Increased sweating |
| S8 | Slow or poor healing of cuts |

## NEUROLOGICAL FUNCTION

**Sensory:**

| P1 | Vision impairment |
|----|---|
| P2 | Hearing impairment |
| P3 | Perception changes taste/smell |
| P4 | Burning sensation |

**Table 2.2** Continued

|  | \multicolumn{6}{c}{Sensory} | \multicolumn{8}{c}{Motor} |
|---|---|---|---|---|---|---|---|---|---|---|---|---|---|---|

| | Sensory | | | | | | Motor | | | | | | | |
|---|---|---|---|---|---|---|---|---|---|---|---|---|---|---|
| | P1 | P2 | P3 | P4 | P5 | P6 | M1 | M2 | M3 | M4 | M5 | M6 | M7 | M8 |
| Hexane | | | | | | | | 5 | | | | | | |
| Hydroquinone | | | | | 5 | | | | | | | | | |
| Kepone | | | | | | 5 | | 5 | 5 | | | | | |
| Lead | 16 | 5 | | 16 | | | | 5 | 5 | 16 | 10 | 16 | 16 | 16 |
| Leptophos | | | | | | | | 5 | | | | | | |
| Manganese | | | | 5 | 5 | | 5 | 5 | | | | | | |
| Mercury | | | | | | | | 10 | | | | | 10 | 10 |
| Methyl Bromide | | | | | | | 5 | | 5 | | | | | |
| Methyl Chloride | | | | | 5 | | 5 | 5 | | | | 14 | 14 | |
| Methyl-n-butyl Ketone | | | | 5 | | | | 5 | | | | | | |
| Methylmercury | | 5 | | 5 | | | 5 | 5 | 5 | | | | | |
| Nitrofurans | | | 3 | | | | | 3 | | | | | | |
| Organo-phosphate Esters | | 5 | | 5 | | | 5 | 5 | | | | | | |
| Organo-phosphates | | | 3 | 3 | | | | 3 | | 3 | | 3 | | 3 |
| PBB | 15 | | 15 | 1 | 15 | | | 15 | | | 15 | 15 | 15 | |
| PCB | | | 6 | 8 | | | | | | | | | 6 | 6 |
| Phenylmercury | | | | | | | 5 | | | | | | | |
| Polychlorinated Polycyclics | 9 | | 9 | | 9 | | | 9 | | | | | | |
| Radiation, ionizing | | | | | | | | | | | | 7 | | |
| Sulfur Dioxide | | 5 | | | | | | 5 | | | | | | |
| Tetraethyl lead | | | | | | | | | | | | | | |
| Tetraethylthiuram Disulphide | | | | 3 | | | | | | | | 3 | | |
| Thallium | | | | | 5 | | | 5 | | | | | | |
| Toluene | | 5 | | | 5 | | | | | | | | | |
| Trichloroethylene | | | | | | | | 5 | | | | | | |

P5    Parasthesias
P6    Hallucinations

**Motor:**

M1    Speech impairment
M2    Muscle weakness
M3    Tremors
M4    Difficulty walking
M5    Seizures

**Table 2.2** Continued

| | NEUROLOGICAL FUNCTION | | |
|---|---|---|---|
| | Associative | | System Arousal |
| | A1 A2 | A3 A4 | A5 | | N1 N2 | N3 N4 | N5 N6 | N7 |
| Acrylamide | | | | | | | | |
| Adiponitrile | | 5 | | | | | | |
| Aniline | | | | | | | | |
| Arsenic | | | | | | | | |
| Arsine | | | | | | | | |
| Benzene | | | | | | 5 5 | | |
| Bromophenylacetylurea | | | | | | | | |
| Carbon Disulfide | 10 10 | | | | 10 | 10 10 | | |
| Carbon Monoxide | | 10 10 | 10 | | | | | |
| Carbon Tetrachloride | | 5 | | | 5 | | | |
| Chlordane | | | | | | | | |
| Chlorinated Hydroquinolines | | | | | | | | |
| Chloroprene | | | | | | 5 5 | | |
| Cyanide | | | | | | | | |
| DDT | | | | | | | 5 | |
| Dichloroethane | | | | | | | | |
| Dimethyl Sulphate | | | | | | | | |
| Dioxin | | | | | | 5 5 | | |
| Ether, diethyl | | | | | | | 5 | |
| Ethylene Dichloride | 5 | | | | | | | |
| Hexachlorophene | | | | | | | | |

M6        Incoordination/clumsiness
M7        Dizziness
M8        Fatigue

**Nervous system arousal:**

N1        Lethargy
N2        Depression
N3        Nervousness
N4        Irritability
N5        Emotional instability

**Table 2.2** Continued

|  | Associative | | | | | System Arousal | | | | | | |
|---|---|---|---|---|---|---|---|---|---|---|---|---|
|  | A1 | A2 | A3 | A4 | A5 | N1 | N2 | N3 | N4 | N5 | N6 | N7 |
| Hexane |  |  |  |  |  |  |  |  |  | 5 |  |  |
| Hydroquinone |  |  |  |  |  |  |  |  |  |  |  |  |
| Kepone |  | 5 | 5 |  |  |  | 5 | 5 | 5 |  |  |  |
| Lead |  | 5 |  |  | 12 |  | 16 | 16 | 10 | 10 |  |  |
| Leptophos |  |  |  |  |  |  |  |  |  |  |  |  |
| Manganese |  | 5 |  |  |  |  |  |  |  |  |  |  |
| Mercury |  | 10 |  |  |  |  | 10 |  | 10 | 10 |  |  |
| Methyl Bromide |  |  | 5 |  |  |  |  |  |  |  |  |  |
| Methyl Chloride |  |  | 5 | 14 |  |  | 14 | 14 |  | 14 |  |  |
| Methyl-n-butyl Ketone |  |  |  |  |  |  |  |  |  |  |  |  |
| Methylmercury |  |  |  |  |  |  |  |  |  |  |  |  |
| Nitrofurans |  |  |  |  |  |  |  |  |  |  |  |  |
| Organo-phosphate Esters |  | 5 |  |  |  |  | 5 | 5 | 5 | 5 |  |  |
| Organo-phosphates | 11 | 11 |  |  |  | 11 |  | 11 |  |  |  |  |
| PBB |  |  |  |  |  |  | 15 | 15 |  |  |  |  |
| PCB |  | 6 |  |  |  |  | 6 | 6 |  |  |  |  |
| Phenylmercury |  |  |  |  |  |  |  |  |  |  |  |  |
| Polychlorinated Polycyclics |  |  |  |  |  | 9 |  |  |  |  |  |  |
| Radiation, ionizing |  |  |  | 7 |  |  |  |  |  |  |  |  |
| Sulfur Dioxide |  |  |  |  |  |  |  |  |  |  |  |  |
| Tetraethyl lead |  |  |  |  |  |  |  |  |  |  |  |  |
| Tetraethylthiuram Disulphide |  |  |  |  |  |  | 3 |  |  |  |  |  |
| Thallium |  |  |  |  | 5 |  | 5 |  |  |  |  |  |
| Toluene |  | 5 | 5 |  |  |  |  |  |  | 5 |  |  |
| Trichloroethylene |  |  |  |  |  |  |  |  |  |  |  |  |

N6      Hyperactivity
N7      Photophobia

**Associative:**

A1      Decreased mental acuity
A2      Impaired memory
A3      Confusion

**Table 2.2** Continued

| | NEURO. FUNCT. Physiological | | | | MUSCULO- SKELETAL | | | GASTROINTESTINAL | | | | | |
|---|---|---|---|---|---|---|---|---|---|---|---|---|---|
| | B1 | B2 | B3 | B4 | K1 | K2 | K3 | G1 | G2 | G3 | G4 | G5 | G6 |
| Acrylamide | | | | | | | | | | | | | |
| Adiponitrile | | | | | | | | | | | | | |
| Aniline | | | | | | | | | | | | | |
| Arsenic | | | | | | | | | | | | | |
| Arsine | | | | | | | | | | | | | |
| Benzene | | | | | | | | | | | | | |
| Bromophenylacetylurea | | | | | | | | | | | | | |
| Carbon Disulfide | 3 | 10 | | | | | | | | | | | |
| Carbon Monoxide | | | | | | | | | | | | | |
| Carbon Tetrachloride | | | | | | | | | | | | | |
| Chlordane | | | | | | | | | | | | | |
| Chlorinated Hydroquinolines | | | | | | | | | | | | | |
| Chloroprene | | | | | | | | | | | | | |
| Cyanide | | | | | | | | | | | | | |
| DDT | | | | | | | | | | | | | |
| Dichloroethane | | | 5 | | | | | | | | | | |
| Dimethyl Sulphate | | | | | | | | | | | | | |
| Dioxin | | | | | | | | | | | | | |
| Ether, diethyl | | | | | | | | | | | | | |
| Ethylene Dichloride | | 5 | | | | | | | | | | | |
| Hexachlorophene | | | | | | | | | | | | | |

A4    Disorientation

A5    Slowed functional adolescent development

**Physiological responses:**

B1    Headaches

B2    Sleeplessness

B3    Sleepiness

B4    Loss of appetite

**MUSCULOSKELETAL**

K1    Joint pain

K2    Swelling in joints

K3    Muscular aches and pains

**Table 2.2** Continued

|  | NEURO. FUNCT. Physiological | | | | MUSCULO- SKELETAL | | | GASTROINTESTINAL | | | | | |
|---|---|---|---|---|---|---|---|---|---|---|---|---|---|
|  | B1 | B2 | B3 | B4 | K1 | K2 | K3 | G1 | G2 | G3 | G4 | G5 | G6 |
| Hexane | | | | | | | | | | | | | |
| Hydroquinone | | | | | | | | | | | | | |
| Kepone | | | | | | | | | | | | | |
| Lead | 16 | 5 | 16 | | | | | | 16 | | | | |
| Leptophos | | | | | | | | | | | | | |
| Manganese | | 5 | | | | | | | | | | | |
| Mercury | 10 | 5 | 10 | | | | | | | | | | |
| Methyl Bromide | | | | | | | | | | | | | |
| Methyl Chloride | | 14 | 14 | 14 | | | | | 14 | | | | |
| Methyl-n-butyl Ketone | | | | | | | | | | | | | |
| Methylmercury | | | | | | | | | | | | | |
| Nitrofurans | | | | | | | | | | | | | |
| Organo-phosphate Esters | | 5 | | | | | | | | | | | |
| Organo-phosphates | | 11 | 11 | | | | 11 | | | | | | |
| PBB | 15 | 15 | 15 | | 1 | 1 | 1 | 1 | 1 | | | 1 | 1 |
| PCB | 6 | 6 | 6 | | 8 | | 8 | 6 | 6 | 8 | 13 | | |
| Phenylmercury | | | | | | | | | | | | | |
| Polychlorinated Polycyclics | 9 | | | | | 9 | | | 9 | | | | 9 |
| Radiation, ionizing | | | | | | | | | | | | | |
| Sulfur Dioxide | | | | | | | | | | | | | |
| Tetraethyl lead | | 5 | | | | | | | | | | | |
| Tetraethylthiuram Disulphide | | | | | | | | | | | | | |
| Thallium | | | | | | | | | | | | | |
| Toluene | | | | | | | | | | | | | |
| Trichloroethylene | | | | | | | | | | | | | |

## GASTROINTESTINAL

| | |
|---|---|
| G1 | 10 lb. or more weight loss |
| G2 | Nausea |
| G3 | Vomiting |
| G4 | Abdominal pain |
| G5 | Abdominal cramps |
| G6 | Diarrhea |

**Table 2.2** Continued

## REFERENCES CITED IN THE TABLE

1. H. A. Anderson et al., "Symptoms and Clinical Abnormalities Following Ingestion of Polybrominated Biphenyl-contaminated Food Products," *Annals of the New York Academy of Science* 320 (1979): 684–702.
2. V. Bencko and K. Symon, "Test of Environmental Exposure to Arsenic and Hearing Changes in Exposed Children," *Environmental Health Perspectives* 19 (1977): 95–101.
3. J. B. Cavanagh, "Peripheral Neuropathy Caused by Chemical Agents," *CRC Critical Reviews in Toxicology* 2, no. 3 (1974): 365.
4. W. E. Dale, A. Curley, and C. Cueto, "Hexane Extractable Chlorinated Insecticides in Human Blood," *Life Sciences* 5 (1966): 47.
5. T. Damstra, "Environmental Chemicals and Nervous System Dysfunction," *Yale Journal of Biology and Medicine* 51 (1978): 457.
6. A. Fischbein et al., "Clinical Findings Among PCB-exposed Capacitor Manufacturing Workers," *Annals of the New York Academy of Science* 320 (1979): 703–15.
7. R. Furchtgott, "Behavioral Effects of Ionizing Radiations," *Psychological Bulletin* 60, no. 2 (1963): 157–99.
8. M. Goto and K. Higuchi, "The Symptomatology of Yusho," *Fukuoka Acta Medicine* 60 (1969): 409–31.
9. R. D. Kimbrough, "The Toxicity of Polychlorinated Polycyclic Compounds and Related Chemicals," *CRC Critical Reviews in Toxicology* 2 (1974): 445–98.
10. N. K. Mello, "Behavioral Toxicology: A Developing Discipline," *Federation Proceedings* 34, no. 9 (1975): 1832–34.
11. D. R. Metcalf and J. H. Holmes, "EEG, Psychological and Neurological Alterations in Humans with Organophosphorus Exposure," *Annals of the New York Academy of Sciences* 160 (1969): 357–65.
12. L. S. Moore and A. I. Fleischman, "Subclinical Lead Toxicity," *Journal of Orthomolecular Psychiatry* 4 (1975): 61–70.
13. H. K. Ouw, G. R. Simpson, and D. S. Siyali, "Use and Health Effects of Aroclor 1242, a Polychlorinated Biphenyl, in an Electrical Industry," *Archives of Environmental Health* 13 (1976): 189–94.
14. J. D. Repko and S. M. Lasley, "Behavioral, Neurological and Toxic Effects of Methyl Chloride," *CRC Critical Reviews in Toxicology* 6, no. 4 (1979): 283.
15. J. A. Valciukas et al., "Comparative Neurobehavioural Study of a Polybrominated Biphenyl-exposed Population in Michigan and a Non-exposed Group in Wisconsin, *Environmental Health Perspectives* 23 (1978): 199–210.
16. J. A. Valciukas et al., "Behavioral Indicators of Lead Neurotoxicity: Results of a Clinical Field Survey," *International Archives of Occupational and Environmental Health* 41 (1978): 217–36.

# 3
# The Limitations of Summary Risk Management Data

## C. RICHARD COTHERN AND DAVID W. SCHNARE

Estimation of the risk due to environmental contaminants involves information about exposure concentrations, the population affected, and the health effects. Using this information to assess risk means using data that is uncertain, and often means using assumptions in place of missing data as well. The range of uncertainties can be estimated when all necessary data is available—what we might call a ponderable situation. In the event where assumptions must be used, our ability to ponder the uncertainty of our risk estimate is seriously limited. Once at this limit, the scientific community has reached the limitation of risk management data. It has no scientific tools by which to ponder the uncertainty of the risk estimates. We call this situation an imponderable one. This chapter takes science outside the laboratory and into the policy arena—at best a most imponderable environment. Therein, one uses scientific judgement to develop a science policy in order to deal with uncertainty. The purpose of this chapter is to examine the leaps in faith that must be made to bridge scientific uncertainties during the estimation of risk. To wit, we ponder the imponderables.

### PONDERABLES AND IMPONDERABLES

Ignorance comes in two kinds of boxes, one with a translucent top and one that is coal black. The former allows us to estimate the nature of and address the degree of our ignorance. For example, we may not know how many people are exposed to carbon tetrachloride in drinking water, but having done a survey on this question, we can bound our estimate of exposure by examining the

**Figure 3.1**
**Contributions to Uncertainty**

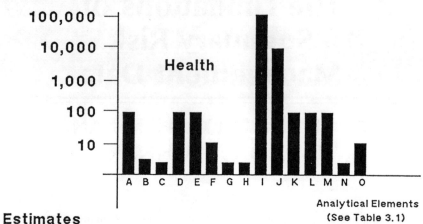

Estimates could be off by a factor of about:

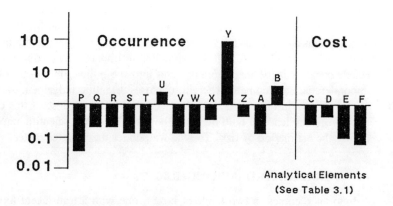

**Table 3.1**
**Uncertainty Contributions to Risk Assessments**

| | Health Assessment | | Occurrence Assessment |
|---|---|---|---|
| a Laboratory Procedures | U* | p Survey Size | U |
| b Tumor Information | U | q Collection Site | O or U |
| c Time-to-Tumor | U | r Collection Time | O or U |
| d Compound Purity | O | s Collection Method | U |
| e Experimental Surroundings | O | t Container Type | U |
| f Use of Upper 95% Confidence Interval | O | u Storage Stability | U |
| g Dietary Considerations | O or U | v Analytical Recovery Percent | U |
| h Tissue Exams | O or U | w Compound Identification | O or U |
| i Synergism/Antagonism | O or U | x Analytical Accuracy | O or U |
| j Sensitive/Average Population | O or U | y Detection Limits | O or U |
| k Animal vs Man | O or U | z Pollutant Level | O |
| l Sample Size | O or U | a Percent Absorption (oral) | O or U |
| m Dose Levels | O or U | b Inhalation Exposure | O or U |
| n Statistical Size | O or U | | |
| o Body Weight vs Surface Area | O or U | Cost Assessment | |
| | | c Unit Costs | O or U |
| | | d Treatment Selection | O or U |
| | | e Industry Growth | O |
| | | f Systems Demand Growth | O |

*U means an underestimate and O an overestimate

**Figure 3.2**
**The Benefits and Costs of Alternative Drinking Water Standards for TCE**

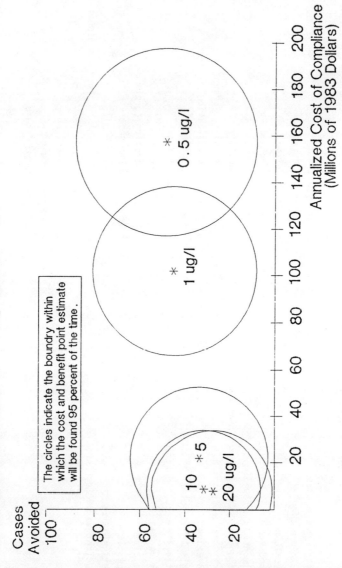

*Source*: U.S. EPA, Regulatory Impact Analysis of Proposed Regulations to Control Volatile Organic Contaminants in Drinking Water, EPA570/9-85-004, Washington, D.C., 1985

## THE LIMITATIONS OF SUMMARY RISK MANAGEMENT DATA 35

probability of the error in our estimate and creating a range of probable exposure. Having done so, we have an estimate of the probable exposure and as well some idea of how certain we are about the accuracy of the estimate. In brief, we can ponder the question of exposure by using available data and adjusting our considerations by recognizing the limitations of the data itself.

The ponderable limitations of risk-management data have been discussed elsewhere and these are, of course, important considerations.[1] To give a feel for the magnitude of uncertainty that is quantifiable, and therefore relatively ponderable, examine Table 3.1 and Figure 3.1. You will note that there is a very large amount of uncertainty in summary risk-management data. In fact, the uncertainty is so large as to suggest that risk managers often cannot differentiate among management options. Figure 3.2 makes this point in an extremely graphic manner. The figure displays the benefits and costs of five different alternatives for setting a drinking water standard for TCE. The uncertainty that surrounds these estimates is shown as a circle around each alternative. Note that the uncertainty in the first three alternatives overwhelms any apparent difference between the point estimates of benefit or cost. If the risk manager was trying to choose between alternatives of 5 and 10 ug/1, this benefit-cost data would be too uncertain to assist him. The lesson here is that even when dealing with the ponderables, it can become almost impossible to distinguish between possible regulatory choices. We shall take up the routinely used solution to this conundrum as we deal with the blacker of our boxes, the imponderables.

Routinely, high school chemistry students are challenged with an early experiment consisting of a sealed shoe box and instructions to take reasonable steps to determine what is within the box without going so far as to open it. While this may be considered a "black box" type problem, it's not as hard as it might be—after all, whatever is within must be smaller than a shoe box. The "black box" confronting the risk manager is certainly a much more difficult analytic problem. It might be akin to telling the student there does exist (somewhere) a black box, but that's all there is known about it, and would the student please take steps to find out what is in this box. Oh, and by the way, would the student please do so by the end of the day, as there is a meeting with the principal at that time, and he wants to know.

Surely, our colleagues have suggested, this is more a question to be dealt with by philosophy students than by chemists. Despite the apparent "scientific" nature of risk management, and the elements that make it up, we tend to agree that when dealing with the imponderable, philosophy may be a more appropriate rubric than the analytical sciences. We now present several thoroughly black boxes and the philosophies that are available to deal with them.

The uncertainties in data gaps involved in risk assessment and the decisions of risk management enter at every level and involve diverse and complex interrelationships and characteristics (see Table 3.2). To aid in the discussion of imponderables, these problem areas or data limitations are divided into four

**Table 3.2**
**Data Limitations and Corresponding Science Policy Options**

| Data limitations | Science policy options | Data limitations | Science policy options |
|---|---|---|---|
| | Health Effects | | |
| No causal link, only correlations | - conservative, assume that the correlation is true | Morbidity versus mortality | - morbidity less important |
| | - caveat that the correlation may not be true | | - balance using cost/benefit |
| | | | - add on an equal basis |
| | - use to substantiate or corroborate but not as a primary finding | | - compare to life shortening |
| | | Hormesis (essential to life) | - disregard health benefit |
| Synergism or antagonism | - assume that neither exists and that effects add linearly | | - use the concept if a break point can be determined |
| | - use a safety factor | Uncertainty (options apply to every category) | - display with point estimates |
| Latent period | - analyze on lifetime basis | | - quantify |
| | - use population risk rather than individual risk | | - list as qualitative effect |
| | | | - use surrogate |
| | - ignore effect | | - worst case approach |
| | | | - choose range like ± 2 <u>orders</u> of magnitude |

| | | | |
|---|---|---|---|
| Health Effects (continued) | | | - add effects linearly |
| | | | - regulate based on worst case |
| | | Exposure | |
| Threshold versus non-threshold (reversibility) | - assume no threshold (conservative) | | - choose most sensitive species |
| | - make choice based on analogy with data for surrogate contaminant | Animal and human variation in uptake, distribution, -relation, excretion, metabolism and sensitivity (conversion from animal to human) | - choose species "closest" to man |
| | - assume that it is a reversible effect and thus acceptable | | - use an average such as reference man for human |
| | - assume biological or practical threshold | | - use safety factors |
| | - reversibility | | - assume man is more sensitive - e.g. factor of 10 |
| All carcinogens are not equal - have different potencies | - weight of evidence | | |
| | - risk assessment | | - use surrogate |
| | - disregard minor effects | | - body weight versus surface area |

37

**Table 3.2** Continued

| Data limitations | Science policy options | Data limitations | Science policy options |
| --- | --- | --- | --- |
| Exposure (continued) | | | |
| Exposure to contaminants is from different media (air, water, food) | - conservative - add worst case<br>- use average or typical concentrations<br>- assume a contribution from each | Unknown contribution from unstudied exposure routes (e.g. dermal) | - allow larger risk for smaller populations<br>- assume it is negligible<br>- safety factor<br>- assume it contributes some arbitrary fraction of risk |
| Only the exposure is known and not what actually got to the cell to do damage | - assume exposure means the system was affected (worst case)<br>- use surrogate to approximate actual dose that got to the organ involved | | - assume or guess at level |
| | | | Analysis of Risk |
| Inestimable exposure due to lower limit to detection ability | - ignore below limit<br>- de minimis (line drawing)- subjective and arbitrary<br>- control to limit of detection or detection efficiency (moving target)<br>- allow larger risk for smaller populations | Animal endpoints may not be found in humans | - assume that an animal cancer signals cancer somewhere in the human<br>- assume will not occur in humans<br>- assume same endpoints |
| | | Possibility that have missed an important endpoint | - assume have correct and most important information identified<br>- safety factor |

## Analysis of Risk (Continued)

| Issue | Approach | Resolution |
|---|---|---|
| Only know toxicity or bioassay at high exposure levels | - ignore acute effects and only consider chronic effects | |
| | - extrapolate into the unknown | |
| | - worst case analysis; flat curve or supralinear | |
| Contaminant may affect other organisms than humans | - base decision on human effect only | |
| | - base decision on environmental effect only | |
| | - choose most sensitive organism | |
| | - choose most sensitive organism of economic value (e.g. Snail Darter as counter example) | |
| | - list all organisms affected | |
| | - limit consideration according to legislative mandate | |
| | No biological basis to select a particular dose-response model | - use multistage model, upper 95% confidence and assume that the dose-response curve is linear at low doses |
| | | - pick a model that appears to be most conservative |
| | | - pick model that gives predictions in middle or towards the most conservative (may not always be the same model) |
| | Uncertainties in risk assessment due to many things including the wide variety of housing, also range of model predictions from extrapolation of dose-response curve, animal conversion by body weight or surface area, experimental uncertainties due to surroundings, chow purity, GLPs and time-to-tumor, representations of water and food | - guess; e.g. try $\pm$ two orders of magnitude |
| | | - worst-case approach |
| | | - use surrogate |
| | | - choice of language-stay away from statistical language |
| | | - qualitative list |
| | | - disregard some errors |
| | | - quantify (statistical data) |

39

**Table 3.2** Continued

| Data limitations | Science policy options | Data limitations | Science policy options |
|---|---|---|---|
| | Other | | |
| Nonexistence of quality assurance information | - assume data is correct<br>- apply uncertainty analysis from a similar measurement<br>- assume protection is provided by using a safety factor | Perception/fear | - cater to perception and set standard at zero<br>- consider it unimportant<br>- claim standard is "safe"<br>- alter perception |
| Complexities such as: natural/man-made, voluntary/involuntary, delayed/immediate, continuous/occasional | - weight by perception or equally (societal view)<br>- impose decision makers values<br>- disregard | Scientists disagree | - try to get consensus<br>- go with majority<br>- go with conservative judgment<br>- go with most respected<br>- do nothing<br>- use political source |
| Determining an acceptable level | - arbitrary choice of individual risk, e.g. 1:1,000,000/lifetime<br>- combine individual risk and population<br>- cost/benefit | Research cost is beyond budget or reason | - limit budget by policy regarding areas to be emphasized<br>- accept current data base and increase assumptions and uncertainty |

arbitrary areas: health effects, exposure, analysis of risk, and others. Each of these areas contains many imponderables and each will be discussed and examined for possible science policy options.

## ASSESSMENT OF HEALTH EFFECTS

### Correlation/Causality

Direct evidence of an adverse health effect in humans is sometimes hard to come by and its cause is even harder to determine. When evidence is available, especially if it indicates there is a human health effect, the most serious and often immediate risk aversion activities are implemented. In some cases, there are unequivocal data documenting a causal link between the purported agent and the resultant harm. For example, there is no question that vinyl chloride at certain levels causes human cancer.[2] For vinyl chloride the mechanism of the disease process is sufficiently well-known that there appears to be no question about causality. For other volatile organic compounds, however, there is some question as to carcinogenic causality.[3] Most human data only indicate a statistical correlation between the assumed health threat and the resultant health effect. In fact, the correlation, while possibly significant, is usually not very large, and only a relatively small amount of the health effect is explained by the assumed threat. An example of the latter is the suggested statistical relationship between breast cancer and radium.[4]

There is no means for quantitating our certainty about causality. Either we assume the link is causal or we don't. A weak science policy middle ground is to assume causality but admit to ourselves that the correlation may be meaningless. However, this does not help the risk assessor or the risk manager. A better solution is to use such correlations only to corroborate a finding that a potential risk is real.

### Interactions

There are a few cases where data indicate that two chemicals when present together pose greater risk than the sum of the risks they present when alone. Asbestos and cigarette smoking is one well-known example.[5] However, this is a very rare case. More typically, there are simply no data on synergy. In like measure, some chemicals are antagonistic with regard to health effects. That is, they pose less risk when present together than they do individually.

The problem here is an absence in underlying biochemical theory upon which to base estimates of synergism or antagonism. Considering the cost of long-term animal studies, there is little reason to believe that direct tests of synergism or antagonism will be undertaken. Such an effort for the nine volatile organic drinking water contaminants the EPA has regulated would require 511 distinct experiments—1,022 if two sexes or species were used, and double again if both were used. This could be doubled again if replication was desired. Therefore, in place of actual animal testing, some biochemical theory must be used, if the basis is to remain scientific. In the absence of a biochemical theory, a policy approach must be used.

Two science policy options are available: the presumption that there is no synergism or antagonism (that effects are additive); or, use of some estimated safety factor to account for possible synergism. Note that the latter case, occasionally used, is especially conservative in that it cannot account for potential antagonism. Of the known synergistic effects, none appear to involve effects larger than a factor of ten and often the effects are less than a factor of two.[6] Thus the magnitude of this uncertainty could be bounded by a science policy.

## Latency Period

For some health-effect endpoints the effect does not occur until several years after the insult. This latent period can range between a few years and a few decades.[7] This time delay makes it difficult to determine an annual risk rate because the latency period may be a significant part of the lifetime.

Because there is a latency period, it would appear important to do analysis of risk on a lifetime basis rather than a per year basis. In this case it is important to use population risk (total expected number of health effects) rather than individual risk (risk per person) in the analysis because it averages the effect over a larger group. Another science policy alternative would be to ignore the effect of the latency period altogether.

## Mortality/Morbidity

Many environmental contaminants produce a plethora of effects among which some are fatal and some are unpleasant but nonfatal. The consequences of this latter morbidity is related to diseases that may be debilitating or have negative outcomes. It is generally accepted that mortality is more important than morbidity, but how much more important is a fatal effect? If the probability of a fatal effect is 100 times less than another negative but not fatal endpoint, are the two on a roughly equal footing?[8] For example, the individual annual risk level of infection for one virus in 10,000 liters of drinking water is of the order of one in ten thousand.[9] However, the risk from volatile organic compounds such as benzene or carbon tetrachloride at some present levels in drinking water is of the order of one in a hundred thousand per lifetime.[10] How can these relative risks be compared?

Generally, the relationship between morbidity and mortality leads to an assumption that morbidity is less important than mortality. Often the relationship can be handled by determining the number of hours lost from sickness compared to lifetime lost through death. These two effects could be balanced using cost/benefit analysis. The third science policy alternative is to add the effects of mortality and morbidity on an equal basis, or fourth, to work out some factor as to their relative importance.

## Hormesis

It is well known that some elements are essential for life; however, there is often some question about the concentration level at which essentiality occurs. This characteristic requirement for life is called hormesis. An example of hormesis is iron. The recommended daily dietary allowance for iron is in the range of 10–20 mg. However, an overdose of 1 to 2 grams of iron may cause death.[11]

When the possibility of hormesis enters an analysis as an imponderable, one can either assume it does not exist or use the concept to determine the level at which, with increasing exposure, the effect changes from being beneficial to detrimental to health.

## Uncertainty

In every measurement some uncertainty is present and the investigation of a bioassay involves many contributions to this uncertainty. Also, variables like species, sex, age, and type of endpoint add to the experimental uncertainties. For example, for many contaminants, mice develop liver cancers while other species show no cancer.[12] The effect of ionizing radiation is not the same for all age groups.[13]

As always, in the analysis of health effects, the ideal situation is to quantify them. However, in very few cases are the data complete enough to allow a complete analysis. Usually, in order to improve the statistical accuracy, different tumor types are added together. Often the categories combined are really incompatible but nonetheless are combined. Examples of such tumor types that could be combined include benign and malignant, male and female test results, and fatal and curable cancers. Technically, each should be treated separately.

Sometimes the uncertainties are so great that all that can be done is to list them qualitatively. In risk analysis one can use a worst-case approach or the data from a surrogate contaminant. Another approach is to choose an arbitrary range of uncertainty, for instance plus or minus two orders of magnitude.

## Thresholds

The effect of some contaminants is reversible below a threshold level while for others there seems to be no apparent threshold. The triggering levels at which contaminants begin to manifest toxic or irreversible effects may not be known or may be obscured by questionable data. It has been suggested that the dose-response curve for radium shows a threshold because no cases have been reported below a relatively high dose.[14] However, the curve appears to be linear going through the origin, so it is possible that an effect can occur at low levels but that not enough cases have been examined to show this.

One of the most difficult issues in health effects is to decide whether or not a threshold exists. The conservative science policy option for this situation is to assume there is no threshold. This assumption is generally made for carcinogens. Two less used but probably equally valid approaches are to make the choice based on an analogy with data for a surrogate chemical, or to assume that the effect is reversible and therefore is an acceptable effect.

## Varying Potencies

It is not surprising that different contaminants have widely varying potencies. For example, EDB is clearly more potent than trichloroethylene.[15] But the lack of knowledge of the mechanism of action makes the determination of the relative potencies difficult. The potency may be different at high exposure levels, but little or nothing may be known about the relative effects at low exposure levels.

Differences in potency become important when only a limited number of exposures can be controlled, perhaps because of cost implications or because control of one requires exposure to the other. Two basic policy approaches are widely in use. Weight of evidence categories can be used to develop a priority scale for different carcinogens, a nonquantitative approach. Alternatively, a full-scale risk assessment would permit comparisons and priority setting. Other science policy options include dealing only with those compounds for which human exposure can be avoided through control or using a worst-case scenario.

## ASSESSMENT OF EXPOSURE

### Animals/Humans

Complexities and uncertainties enter exposure analysis in a number of areas. One example is the wide variation among different animal species and between animals and humans in their uptake, distribution, retention, excretion, metabolism, and sensitivity.[16] The animal uptake of radium is of the order of a few percent (1–3 percent), while that of humans is of the order of 20 percent.[17] In converting bioassay data relating animals to humans, these various differences must be taken into account. Unfortunately, correlations are not documented for most chemical contaminants.

A number of science policy options exist for dealing with the complexities due to differences between animal species and humans. One policy is to choose the most sensitive species. Another is to choose the species that is "closest" to man. In some cases enough data exist that an average value is available such as the Reference Man for humans.[18] One could assume that man is more sensitive than animals by, for example, a factor of ten. In the comparison between species or in the conversion of bioassay data from animals to humans, the problem exists

THE LIMITATIONS OF SUMMARY RISK MANAGEMENT DATA    45

of how to compensate for the different size. The common factors used in this extrapolation are to relate the body weight directly or to use the ratio of body surface areas to body weights.

## Transport

For environmental contaminants, the largest contribution to the overall uncertainty usually occurs in the area of how the contaminant gets from the source to the actual cells affected. This is usually described by the terms "fate" and "transport." One has to estimate how much material evolves from a source and how well it moves through the environment. This may involve movement through the air, water, soil, or the food chain. Often the complexities of the different alternative pathways lead to large uncertainties in the estimate of how the contaminant moves from the source to the person involved. Radioactive thorium is highly insoluble in water while its progeny, radium, is much more soluble.[19] Thus, radium can reach the drinking water tap, while thorium is unlikely to get that far.

In combining exposures from the air, water, and food pathways one could use conservative science policy and add the worst-case values. A middle ground would be to add to the average values for the contribution of each. An alternative is to assume a fixed contribution from each, for example 20 percent from drinking water.

## Exposure/Metabolite

One important recently considered area of exposure is the determination of the kind and amount of the contaminant that actually gets to the cell to do the damage. It is often very misleading and confusing to determine the level of exposure to an organism externally and not to determine how much of the product causing the damage gets to the internal point of damage.[20]

Usually only the exposure to the organism is known and not the actual exposure to the cell where the damage is done. The conservative science policy approach is to assume that all of the contaminant got to the system. This is not necessarily the worst case because the contaminant could be metabolized into something that is more toxic. Another science policy choice is to use a surrogate to approximate the actual dose that got to the cellular level.

## Measurement Parameters

Data describing the concentration level of contaminants in the environment are often limited by the ability to detect their presence. As the lower level of analytical detection is approached, the uncertainty in the measurements becomes larger. Also of importance are the input parameters and physiological characteristics of the organism being considered. For example, the U.S. Environmental

Protection Agency assumes that the intake level of drinking water is two liters per day and that the average person inhales twenty cubic meters of air per day. These values are considered by some to be extreme and by others to be average values. Another complexity is the routes by which a contaminant from the environment can get into the body. These involve the contributions from the inhalation, ingestion, and dermal routes of exposure. In most cases the magnitude of the contribution from the dermal route of exposure is not known.

Thus, in determining the exposure contribution to the estimation of risk, there is a limitation because of the lower limit of detection, the upper limit to treatment efficiency, and the complexities such as route of exposure and intake parameters. There may possibly be a lower limit. This is not zero but may be down in what the engineers call the "noise." Science policy options to deal with this situation include determining a de minimis level or a kind of line-drawing exercise. This approach is subjective and, like most of the philosophic fixes discussed in this chapter, has a certain arbitrary characteristic about it. It is this arbitrary approach that has raised the ire of environmental lobbyists. However, it is no more arbitrary than the assumption of no threshold for carcinogens.

Another approach defines zero as the limit of detection or detection efficiency. The disadvantage of this approach is that it is a moving target since as time goes on the detection efficiency improves. The de minimis level can be a function of population size. The larger the population involved, the lower the acceptable individual risk.

## ANALYSIS OF RISK

### No Corresponding Human Endpoint

In performing bioassay experiments, one of the problems that emerges is the possibility that in the analysis one of the important endpoints has been missed. In addition, if another endpoint that was unexpected occurs, it may well also be missed. The number of health effects due to exposure to lead has required several experimental studies to define completely.

To add to the confusion, the health endpoints for animals may not be the same as those for human beings (for example, effects on the zymbal gland that humans don't have, although there appears to be a correlated toxic impact on the liver). The science policy options in this situation are first to assume that if a cancer exists in an animal, the same cancer will exist in humans. Another approach is to assume that although the endpoint may not be the same in humans, if there is a cancer endpoint in animals there will be a cancer endpoint for some organ in human beings. One could also assume that it does not occur in humans.

### Acute/Chronic

In almost every case one only knows the toxicity or bioassay information at high levels of exposure. There is no guarantee that these effects scale down for low-level exposure in a systematic manner. Often these are acute effects and they mask what may be more serious chronic effects. The phenomenon of cell killing is often added artificially by introducing an exponential term in the dose-response analytical expression.[21]

One possible science policy here is to ignore the acute effects and only consider the chronic effects. Another policy option is to assume that it is possible to extrapolate from the known effects at high levels to those unknown effects at low levels. As always, it's possible to assume a worst-case kind of analysis that either could be a flat linear dose-response curve or to assume that the curve is supralinear.

### Human/Environmental Effect

Considering the environment as a whole, a contaminant may affect organisms other than human beings (for example, DDT and its effect on the peregrine falcon). There are living things in the water that often are more sensitive to contaminants than human beings. Possible science-policy options for this situation are to base the decision on the human effect only or to base the decision on the environmental effect only. The choice of one or the other must reflect the values of the decision maker, and are arbitrary, although not necessarily capricious.

## OTHER IMPONDERABLES

Besides the imponderables mentioned above that are directly a part of the risk-estimation process, there are others that bear some relation to that process but don't have a direct quantitative contribution. These important areas include quality assurance, complexities, perception, budgets, disagreement among scientists, and the general question of what is a safe or acceptable risk level.

### Quality Assurance

For much of the data used in estimating the concentrations of contaminants in the environment, quality-assurance data is nonexistent. Only in a few cases are blind, spiked, and split samples used in the analysis to verify the performance level of the laboratory involved. Thus, there is a contribution to the uncertainty that could be quite large and is, in general, unknown. Often this uncertainty is not random but is in the form of a bias to higher or lower values. Recent audits of the bioassay studies of vinylidene chloride and methyl chloroform have shown problems due to the lack of a quality-assurance program.

There are a number of science policy options that can be used to handle the situation when no quality-assurance data exists. The first is to assume that the data and the uncertainties listed are correct and there are no biases involved. The second approach would be to apply some kind of uncertainty analysis to compensate for the lack of quality-assurance data. The third option would be to apply a safety factor to compensate for any unknown biases.

### Complexities

There are a number of complexities or elements that enter into the assessment of risk. Examples of these are the difference between voluntary and involuntary, natural and man-made, delayed and immediate, continuous and occasional, and others. Involved in many of these elements is the characteristic of perception by the public and often a psychological fear such as that associated with cancer or radioactivity.[22]

Science policy options to deal with complexities pose some difficulties. However, the complexities listed above have been generally quantified as to their relative importance. For example, risks due to exposure to naturally occurring contaminants are accepted about twenty times more readily than those due to man-made sources. Risks taken voluntarily are about a hundred times more acceptable than risks taken involuntarily.[23]

One could weigh these different risks according to the perception of the public or they could be weighed equally as possible science policy options. An alternative approach is to assume that one of these is acceptable or more acceptable than the other. Using these complexities forces the regulatory standards to lower values, and in some cases the perception is so strong that the regulatory standard has to be set almost at zero if not actually at zero. This again raises the whole problem of defining zero in terms of measurement or treatment.

### Acceptable Level

In order to determine what level should cause action after a risk analysis, the question of what is an acceptable level often is asked. An acceptable level would be determined in terms of population risk, individual risk, and some of the elements described in the previous paragraph. Whatever level is chosen as acceptable, it is usually an arbitrary one and always involves subjective decisions. In this area and in many others, scientists often disagree and have genuine differences of opinion. These differences often reflect the uncertainties and complexities involved in analyzing risk.

Determining an acceptable level involves science policy options such as choosing an arbitrary level of $10^{-4}$ health effects per lifetime or maybe a lower level for carcinogenic risk. Another science policy option would be to combine the individual risk and the population in a similar analysis. Another alternative is to use cost/benefit analysis. This latter often reflects limited budgets and limited available funds.

## Perception/Fear

The thought of being exposed to a contaminant that may cause cancer, especially one that we cannot sense (like radiation) leads to perceptions based on fear. The individual risk may be quite low (for example, one in a million) but the fear of this risk often leaves people with the sense that the risk is much larger than it actually is.

In response to this, one science policy option would suggest that the government cater to the perception and set the standard at zero as was done by the Delaney Clause by the FDA and by the Maximum Contaminant Level Goals (MCLGs) for carcinogens under the Safe Drinking Water Act. An alternative science policy would be to ignore the fear or psychological element and set the standard according to the same rules and procedures used for other contaminants.

## Scientists Disagree

Because of the many problem areas and imponderables discussed here, there rarely is agreement within the scientific community on any given issue. For example, there are even strong disagreements on which contaminants are carcinogens, cocarcinogens, promoters, etc. Science policy options in this area involve trying to obtain a consensus, going with the majority, going with the conservative judgment, going with the most respected opinion, or doing nothing.

## Research Cost

Limited resources are available to do research on particular contaminants or to investigate generic questions. Although this problem may be more serious at particular times than others, it is always there. In the absence of research, many risks go unanalyzed, making them part and parcel of the imponderable set of risk-assessment problems. The usual approach to this problem is to make policy decisions regarding which areas are to be emphasized. The alternative would be to accept the present situation and simply expand the uncertainty bounds when questions arise.

## CONCLUSION

This chapter has focused on the imponderables and limitations in summary risk-management data and has presented some options concerning how to manage them to reach regulatory decisions. The approach to dealing with the imponderables is to determine the science policy options and choose one. These options have their base in science and technology but the choice is one of risk management policy. In many cases the option can be clearly stated and thus the risk management process is clear and understandable although differences of opinion will arise.

One of the major problems in the interpretation and use of risk assessment is the combination of the uncertainties involved in the known information and data and those due to the imponderables. The range of estimates developed numerically from information about the uncertainties in the ponderables can be quite large. For example, the estimated population risks for some volatile organic compounds in drinking water have a range of uncertainty from four to six orders of magnitude as shown in Figure 3.2.[24] However, the uncertainty due to the imponderables discussed here will undoubtedly add much to that already known to be due to the ponderables. Thus, we are too often operating in the dark—and may continue to do so for some time.

It may well be an advantage that the human brain is required to integrate the range of the information concerning the ponderables and the science policy options for the imponderables. The human brain is still more capable of coping with uncertainty than a computer unless the data are very well known. The gut feeling of the scientist about the quality of the data, information, or estimates can be as good as or even better than the analysis of the computer.

If the government is to act upon risk assessment data and make science policy choices, its decisions will have an arbitrary character because of the many imponderables. However, since the decisions are based in part on scientific data, they need not be capricious. The suggested policies discussed here are consistent, albeit potentially arbitrary.

It is the responsibility of the risk manager to make decisions in the face of large uncertainties in the ponderables and the range of imponderables discussed here. Regardless of the arbitrariness of such decisions, they can create confidence in society that something is being done to protect public health.

A scientist will tell you what he observed, may provide you a theory within which to place the observation, but cannot honestly promise that the theories are correct. Unfortunately, in the risk management arena, there is a need to act.

Under these conditions, and considering the many imponderables that contribute to the uncertainty in risk management, we must conclude by stating unequivocally that no one can certify the validity of policy approaches available to deal with imponderables. Much like a sailor without his charts, once facts are left behind, the risk manager must operate on his own volition. The solution may be to choose good managers rather than develop further theoretical models or arguments.

The risk management decision process may seem like a shot in the dark. But it may well be better to make decisions with little or no information than to not make such decisions.

When it comes to black boxes, even the most senior risk manager is nothing more than a student who is trying to determine what is inside. In the case of risk management, it is not the principal but society who needs the answer.

# NOTES

The thoughts and ideas expressed in this chapter are those of the authors and are not necessarily those of the U.S. Environmental Protection Agency. This chapter was published in part under the same title in *Drug Metabolism Reviews* 17 (1986):145–69 and is reprinted with the permission of Marcel Dekker, Inc.

1. C. R. Cothern, W. A. Coniglio, and W. L. Marcus, *Techniques for the Assessment of Carcinogenic Risk to the U.S. Population Due to Exposure From Selected Volatile Organic Compounds from Drinking Water via the Ingestion, Inhalation, and Dermal Routes*, U.S. Environmental Protection Agency, Office of Drinking Water (WH–550), EPA 570/9–85–001 (Washington, D.C., 1985).
2. V. J. Feron et al., "Lifespan Oral Toxicity Study of Vinyl Chloride in Rats," *Food and Cosmetic Toxicology* 19 (1981):317–31.
3. National Academy of Sciences, *Drinking Water and Health*, vol. 1 (Washington, D.C., 1982). Also see volumes 2–5.
4. J. A. Bean et al., "Drinking Water and Cancer in Iowa," *American Journal of Epidemiology*, 116 (1982):912–32.
5. National Institutes of Health, *Report of the National Institutes of Health Ad Hoc Working Group to Develop Radioepidemiological Tables*, NIH Publication No. 85–2748 (Washington, D.C.: GPO, 1985).
6. H. F. Smyth, "An Exploration of Joint Toxic Actions: 27 Industrial Chemicals Intubated in Rats in All Possible Pairs," *Toxicology and Applied Pharmacology* 14 (1969):340–47.
7. National Academy of Sciences, *The Effects on Populations of Exposure to Low Levels of Ionizing Radiation* (BEIR III Report) (Washington, D.C., 1980). See also, NAS, *Drinking Water and Health*.
8. D. Latai, D. D. Lanning, and N. R. Rasmussen, "The Public Perception of Risk," in *The Analysis of Actual and Perceived Risks*, ed. V. T. Covello, W. G. Flamm, J. V. Rodericks, and R. G. Tardiff (New York: Plenum Press, 1983).
9. C. P. Gerba, "Strategies for the Control of Viruses in Drinking Water" (Unpublished report, Department of Microbiology, University of Arizona, Tucson, Ariz.).
10. Cothern, *Techniques for the Assessment of Carcinogenic Risk*.
11. A. G. Gilman, L. S. Goodman, and A. Gilman, *The Pharmacological Basis of Therapeutics* (New York: Macmillan, 1980).
12. Nutrition Foundation, *The Relevance of Mouse Liver Hepatoma to Human Carcinogenic Risk, A Report of the International Expert Advisory Committee to the Nutrition Foundation* (Washington, D.C., 1983).
13. NIH, *Report of the National Institutes of Health*.
14. R. D. Evans, "Radium in Man," *Health Physics* 27 (1974):497.
15. U.S. Environmental Protection Agency, *Health Assessment Document for Carbon Tetrachloride*, EPA–600/8–82–001F (Cincinnati, Ohio: Environmental Criteria and Assessment Office, 1985).
16. E. Crouch and R. Wilson, "Interspecies Comparison of Carcinogenic Potency," *Journal of Toxicology and Environmental Health* 5 (1979):1095–1118.
17. J. F. Stara et al., "Comparative Metabolism of Radionuclides in Animals: A Review," *Health Physics* 20 (1971):113–37.
18. International Commission on Radiological Protection (ICRP), *Report of the Task Group on Reference Man*, no. 23 (New York: Pergamon Press, 1981).

19. C. T. Hess et al., "The Occurrence of Radioactivity in Public Water Supplies in the United States," *Health Physics*, 48 (1985):553–86.
20. Ibid.
21. NAS, *The Effects of Low Levels of Ionizing Radiation*.
22. Latai, "The Public Perception of Risk."
23. Ibid.
24. Cothern, *Techniques for the Assessment of Carcinogenic Risk*.

# 4

# First Do No Harm: Diagnosis and Treatment of the Chemically Exposed

## DAVID E. ROOT AND DAVID W. SCHNARE

**HUMAN CONTAMINATION**

Concern about the storage of hazardous chemicals within the human body dates from as early as the fifteenth century. Jesuits required workers in their cinnabar (mercury) mines to receive detoxification treatment when they were no longer able to write their name evenly on lined paper.[1] From that time forward, exposure to chemicals has increased with an ever-quickening pace. The chemical revolution that took place after World War II dramatically increased the number and variety of hazardous chemicals that will store in human tissues. As of 1980 over 400 chemicals have been identified in human tissue, some 48 in adipose tissue (fat).[2]

The nature of this human contamination takes two forms. There are exposures like the one in Michigan where people were contaminated through the dairy and beef food chain.[3] These types of exposures affect large populations and have occurred in many nations. Their stories are legend in the environmental community and include dioxin in Seveso, Italy, PCBs in Japan and Taiwan, methyl isocyanate in Bhopal, India, and HCBs in Turkey. Taken together, over 10 million people have been exposed around the world by these five incidents alone.

The second form of exposure is more pedestrian. In everyday life people are exposed to low levels of chemicals at every turn. Whether it's due to a leaking electrical transformer containing PCBs, gasoline fumes containing benzene, or vegetable juice in steel cans contaminated by lead solder, it's impossible to avoid contamination. Not even infants are immune. They receive a chemical legacy from their mothers during gestation as well as while feeding at the breast.[4] The practical result of these routine low-level exposures is increases in human body burdens of environmental chemicals with age because it is easier for many chemicals to accumulate in bone and fat than to be excreted from the body.

The degree of chemical exposure is difficult to overlook. For example, despite the fact that DDT was banned from production and use in the United States over ten years ago, the percentage of the population having DDT residuals in their bodies continued to grow during the last decade. Today over 99 percent of all U.S. residents have measurable levels of DDT (or its metabolite DDE) in their fat. Similar findings document human body burdens of dieldrin, heptachlor, heptachlor epoxide, and PCBs.[5]

The magnitude of this human contamination, while large, is certainly underestimated. There is no standard analytical method available to determine fat levels of diazepam, cocaine, or phencyclidene (the street drug called PCP or angel dust). There isn't even a standard method for fat analysis of the major active chemical in marijuana (tetrahydrocannabinol [THC]), or its metabolites. Over 50 million people in the nation are likely to have body burdens of these fat-soluble drugs.

## THE SIGNIFICANCE OF HUMAN CONTAMINATION

We must echo the concerns of Rene Dubos, who noted over fifteen years ago: "The greatest danger of pollution may well be that we shall tolerate levels of it so low as to have no acute nuisance value, but sufficiently high, nevertheless, to cause delayed pathological effects and despoil the quality of life."[6] It appears that this condition has now come to pass.

If harmful, fat-soluble chemicals moved into the major fat stores of the body (the adipose tissue) and remained in place, perhaps there would be little need to worry about human body burdens. However, these chemicals do not stay put, nor do they go only to relatively unimportant fat deposits. Whenever lipids (fats) move into the blood, so too do the chemicals stored in them.[7] This occurs every day as part of the normal functioning of the body. For example, the evening fast (while sleeping at night), aerobic exercise, and common emotional stress mobilize fat and hence stored chemicals.[8]

Once in the blood these chemicals have the opportunity to reach every part of the body. For some chemicals this means they will be broken down into components, for example, by reaction in the liver. The components may go back to the fat or, if they are water soluble, may be excreted. However, many of the chemicals industry has created do not easily break down. They were developed so that they would not break down. A good example are the polychlorinated biphenyls (PCBs). These chemicals were made for use in high-temperature environments and even withstand moderately hot fires.

The body is equipped to excrete water-soluble chemicals, but is not as well developed to excrete the fat-soluble ones. Therefore, if the body can't break

DIAGNOSIS AND TREATMENT OF THE CHEMICALLY EXPOSED    55

these chemicals down, they tend to redistribute into the various fatty portions of the body. Besides the adipose tissue lying just below the skin layers of the body, the brain, the sheathing of the nervous system, and the liver are also major fat depots. Persistent human contaminants are routinely found in these depots.[9]

The effect of chemicals that store in the body is not easy to describe. First, as noted in Chapter 2, there is little data on the health effects of chemicals. What data does exist tends to reflect effects in rodents rather than humans, and is nearly always the result of large exposures to a single chemical. It is difficult, if not impossible, to estimate the effects of chemical mixtures, and there have been very few such tests in animals. In addition, animal tests are notoriously poor at indicating the potential for effect on the skin or the nervous system.

The basis for most human data comes from study of populations exposed in the workplace or in large-scale exposures such as those mentioned in the opening paragraphs of this chapter. There are usually a wide number of effects; some may be unique to the chemicals in question, but most certainly are not.

## PHASED DIAGNOSIS

### The Challenge

Diagnosis of chemical intoxication has traditionally required a history of exposure to a known chemical and the finding of appropriate symptoms and clinical signs. Estimations of health effects, as discussed in Chapter 2, are based on the known effects of exposure to one chemical at a time over a known or limited time period. The result is typically estimation of an "acceptable daily intake," based on a "no effect level" and "safety factors." These and many similar measures have been considered sacrosanct for so long that the weakness of their rationale may come as a shock to most health scientists.[10] Further, most of these measures are based on pathological and tissue findings where possible, rather than on the precursor symptoms and signs known to be associated with chemical intoxication.[11]

One weakness in the diagnostic approach still being taught in medical schools with weak occupational or environmental medicine programs is that the approach is insensitive to cumulative effects. Because many chemicals affect the human system in similar ways, it may be that only a small exposure is necessary to push the organism over the threshold at which clear symptoms, and eventually clinical signs, arise. Yet the traditional toxicological approach would suggest that a higher dosage is required.

Another weakness is lack of recognition that health effects associated with chemical exposure are characterized by a hierarchy of events, as discussed in Chapter 2. This hierarchy of effect severity is continuous up to the point of death. Traditional medical models define morbidities by clinical and subclinical signs. These are changes at the system level, the organ level, or the cellular level. They are generally measurable quantitatively, but may be directly observable by a clinician in a qualitative manner. Examples of these adverse effects which can be associated with chemical body burdens include increased blood pressure, decreases in numbers of red blood cells, increases in white blood cells, increases in specific enzymes, decreases in the rate of nerve impulse transmission, rashes, and IQ and personality trait changes.

The more subtle effects of chemical exposure have been defined through "subliminal toxicology."[12] These adverse effects are observed as subtle functional changes such as slowing of motor reactions, impaired regulation of appetite, reduced visual discrimination capacities, fatigue, and memory loss. They are characterized by vagueness and ambiguity.[13] These symptoms constitute the most difficult of all diagnostic challenges. In fact as Weiss points out, "these are not deficits that induce people to seek out physicians."[14]

The significance of chemical exposure is not that low-level exposures can cause subtle symptoms. Rather it is that such symptoms are the sentinels of more serious chemical-related disease. Underlying toxicology research is the time-tested model of biological action: the greater the chemical exposure, the greater the resultant effect. This model has been found to hold true for chemical health threats such as those now found in environmental and occupational settings. In the most common cases of chronic intoxication, all clinical tests are usually negative, despite the clear and plainly undesirable symptomatology.[15] However, as body burdens rise, so too do concentrations in the blood, and resulting exposures in vital organ systems. It comes as little surprise, therefore, to find a progression to more serious disease states with increasing body burdens, as shown in the polybrominated biphenyl example in Chapter 2.

## A Practical Approach

An evaluation schema for patient referrals that accounts for the pathologic factors especially problematic with environmentally persistent chemical contaminants can be described as a five-step process.

*Phase I: Initial Evaluation.* An in-depth investigation is undertaken to describe a patient's chemical exposure and medical history, physical examination findings, laboratory findings, psychological test findings, and any other data pertinent to an initial diagnosis. This could include blood analysis, evaluation of the immune system response of the individual, assessment of the coping mechanisms of the patient, and other behavioral indicators of neurological compromise.

While any one chemical exposure could have been the proximate cause of a health effect, it is much more likely that the latest exposure was simply the overwhelming dose whose effect would not have been felt without multiple earlier exposures, many of which might be quite small in nature. It is critical that the initial evaluation seek information on any earlier exposures that may have come from occupational, hobby, or apparently unrelated activities, but which might, nonetheless, constitute a significant chemical burden. A history of such exposure naturally tilts the scale toward a diagnosis of chemical etiology. It is surprising how common such earlier exposures are.

*Phase II: Initial Diagnosis.* Development of a relationship between the exposure history, medical history, symptoms, and signs requires familiarity with the environmental medicine and toxicology literature, and may often require a consult with a toxicologist or occupational specialist who has familiarity with the various environmental chemicals to which the patient was exposed. The complexity of such a correlation is demonstrated in Chapter 2 where several chemicals appear to cause similar symptoms, symptoms that might also be caused by an organic disfunction associated with anything from bacterial or viral infection to genetic disorders.

It is possible to ascribe the patient's condition to a variety of etiologies, including psychological problems. Our experience suggests that it is simpler, cheaper, and less invasive to the whole patient to rule out organic and chemical etiologies before considering a course of psychiatric treatment. This is done through the next three phases.

*Phase III: Adipose Tissue Sampling and Initial Treatment.* Initial diagnosis of chemical-exposure-related symptomotology is typically verifiable through (a)analysis of chemical body burdens, followed by (b)treatment to reduce such body burdens and to provide relief from reversible burden-related symptoms.

When evaluating the chemical body burden data, it is important to examine the total burden. While a single high peak may be important, it is no less important than the presence of many chemicals whose total burden is significantly above the limits of detection.

Recall, as well, that there are limits to what can be analyzed in human adipose tissue. For example, a person with a significant history of drug abuse might fall prey to the effects of an environmental exposure more quickly, due to the burden of psychoactive drugs he also carries in his system, although methods to detect drugs of abuse in fat are quite limited. Balance previous known exposures against what can be measured in fat. Look specifically for the chemicals known to cause the symptoms being presented by the patient—there may be several. Look to verify that the patient is carrying a burden of the chemical he asserts he has been exposed to. The half-life of most chemicals, including the water-soluble metabolites of tetrahydrocannabinol (for example), is usually measured in months and years. If there is an analytical method for the chemical of interest, and there was exposure, it is likely that it will be found in the fat. If it cannot be found, either the analytical measure is not adequate or no significant exposure took place.

In the event that there is no analytical method available to verify an exposure, it may be sufficient to verify the presence of measurable contaminants, since the synergistic effect of multiple exposures is likely to be the most important aspect of the individual's exposure history in the first place.

It also pays to recognize that the emotional trauma of a chemical exposure can initiate a significant mobilization of preexisting contaminants from fat stores.

Hence, while a new exposure might be small, the mobilization due to the stress of the event may be significant. In most of these cases, symptoms are not as intense and are more sporadic. Nevertheless, they too constitute a result of society's fear of living in a chemical environment, and are properly considered a bonafide disease state brought about by a chemical exposure.

One form of noninvasive chemical body burden reduction treatment is briefly discussed later in this chapter, and use of it has constituted the integration of treatment as part of the diagnostic process. The purpose of such treatment is to reduce chemical body burdens. If reduction of chemical body burdens has an attendant positive effect on symptoms and signs, then there is a strong suggestion that a chemical exposure was the etiology of the disease state.

*Phase IV: Immediate Posttreatment Evaluation.* The purpose of this evaluation is to verify any body burden reduction and symptom remission. It may be advisable to conduct only those clinical tests that would be expected to have reached an equilibrium state within hours, although evaluation of symptoms and some neurological conditions deserve immediate attention.

*Phase V: Follow-up Evaluation.* Routine follow-up consists of (a) a third fat biopsy to verify the equilibrium burden reductions due to treatment, (b) reevaluation of status of symptoms, and (c) recording of any further chemical exposure that may have a bearing on symptom remission and follow-up burden analysis.

The purpose of this follow-up evaluation is to determine the need for more extensive diagnostic procedures such as NMR (for organic disease), neurological work-up, or psychiatric evaluation. These additional high-cost, invasive diagnostics are indicated if body burdens were reduced but symptom remission was not stable, especially where chemical reexposure was not significant. The reasons these additional diagnostics would be necessary would be to determine the cause of the symptoms since, under conditions of reduced body burden, it is unlikely that the continuing symptoms are related to the chemical exposure, unless the symptoms are of the sort that are irreversible. Typically, for low-level contamination incidents, only long-standing neurological symptoms of significant debilitating nature have been found to be irreversible.

Commencement of additional diagnostics should await the follow-up adipose tissue analysis, should the post-treatment analysis not indicate a significant reduction in body burden. Interestingly, body burdens can be significantly lower at follow-up than at immediate post-treatment, and symptoms normally associated with chemical exposure can be found in remission at follow-up, as they may be immediately post-treatment in the event that the treatment was immediately effective.

## TREATMENT OF CHEMICALLY EXPOSED PATIENTS

Treatment of the chemically exposed must be split into an acute and chronic response, the latter of which is most salient to the discussion in this book. Nonetheless, it pays to consider acute treatment.

When a patient presents himself for emergency treatment with the complaint that he has just been exposed to a chemical, the attention of the physician should be on minimizing transfer of the chemical into the bloodstream. For dermal exposures this typically means cleansing the skin and changing clothes. It may also mean leaving an exposure-laden environment. For ingestion this typically means ingestion of an agent that will preclude transfer of the chemical through the gut wall. Examples of such agents include activated carbon, sucrose polyester, and cholestyramine.

These responses, however, are for the massively exposed. The more typical exposure is chronic in nature, taking place over weeks, months, and years. In that case, a different treatment is necessary.

Reduction of fat-stored body burdens requires two basic steps: residue mobilization and enhanced excretion from the bloodstream and the body. The antecedents of body burden reduction research have narrowed active work to treatments that enhance excretion of chemicals through the bile and feces by ingestion of paraffin, activated carbon, or saturated and unsaturated oils.[16] The greatest successes have been with the unsaturated oils and have led to human participation in reduction studies. However, treatments that do not also ensure mobilization from deep stores in fatty compartments will not be efficient.

The key to enhanced excretion lies in overcoming enterohepatic recirculation. While cholestyramine, high-fiber diets, vegetable diets, sucrose polyester, and paraffin have all been used with varying degrees of success, only polyunsaturated oil has significantly enhanced excretion of extremely persistent chemicals and at the same time not increased fat deposition in the liver.[17] Associated with overcoming enterohepatic recirculation is the potential for reduced absorption of important nutrients and thus increased toxicity of persistent chemicals such as PBB. In such cases, increased administration of nutrients, including fat-soluble vitamins, has been found to provide protection in the face of expected toxicity.[18]

As mentioned, enhanced excretion is inefficient without enhanced mobilization of fat-stored moieties, as enhanced excretion merely leaches the blood burden but does little for the burden residing in the fat. Since the concentration ratio of blood to fat is typically about 1:500 for environmental chemicals, mobilization from the fat is critical to reduced blood levels and reduced "internal" exposures.

The active fraction of the adipose tissue constitutes only 5 percent of the fat deposit and does not appear to contain many contaminants found in the inactive fraction.[19] However, it is clear that contaminants found in the deeper fraction are regularly mobilized.[20] While knowledge in this area is relatively poor, mobilization of fat-stored chemicals in the absence of enhanced excretion pathways has been reported to cause latent exposure crises such as hallucinogenic "flashback" events which have kept occupationally drug-exposed police officers off the work force.[21]

Treatments that enhance both mobilization and excretion are few in number. An example of recent research will show the degree of assistance that can be provided using existing practices.

A treatment regimen we have reported on elsewhere is a relatively complex three-week regimen of polyunsaturated oil supplement, heat stress, and vitamin and mineral supplements.[22] Study of four treatment populations documents that body burden reduction and remission of typical chemical symptomotology can be achieved.[23]

Comparing symptom prevalence of chemically exposed and unexposed reference populations with a chemically exposed treatment group shows the effect of reducing body burdens in symptomatic patients (Table 4.1).

Documentation of burden reduction is given in Table 4.2, which shows the percent reductions in adipose tissue concentrations in PBB-exposed individuals receiving treatment. Body burden reductions averaged over 40 percent, and mean burdens of all sixteen chemicals studied were found at reduced levels.

The significance of increasing chemical body burdens in individuals or the population at large comes from the increased probability of chronic disease, whether subtle or acutely manifest. To reduce body burdens is to increase the period before which acute manifestations of disease present themselves. In brief, to reduce body burdens is to put off previously uninitiated, chronic, chemical-related disease.

Because the course of diagnostics and first-step treatment are relatively low in cost, usually less than $3,000, there is some logic in recommending low-cost settlements, or at least initial settlements to the chemically exposed seeking relief under workers' compensation. In the event liability is admitted by the proximate source of the chemical exposure, this too may be a cheap first step that might reduce overall liability, including the cost of long-term litigation.

## NOTES

The thoughts and ideas expressed in this chapter are those of the authors and are not necessarily those of the U.S. Environmental Protection Agency.

1. W. Stopford, "Industrial Exposure to Mercury," in *The Biogeochemistry of Mercury in the Environment*, ed. J. O. Nriagu (New York: Elsevier/North-Holland Biomedical Press, 1979), p. 383.

2. U.S. Environmental Protection Agency, *Chemicals Identified in Human Biological Media, a Data Base*, EPA 560/13–80–036B, PB81–161–176 (Washington, D.C., 1980).

3. H. S. Anderson et al., "Symptoms and Clinical Abnormalities following Ingestion of Polybrominated Biphenyl-contaminated Food Produce," *Annals of the New York Academy of Science* 320 (1979):684–702.

4. R. Lucas, V. Iannachione, and D. Melroy, *PCBs in Human Adipose Tissue and Mother's Milk*, Research Triangle Park Report RTP/1864/50–03F, November 1982.

5. F. W. Kutz, S. Strassman, and J. Sperling, "Survey of Selected Organochlorine Pesticides in the General Population of the United States: Fiscal Years 1970–1975," *Annals of the New York Academy of Science* 320 (1979):60–68.

6. R. Dubos, "Adapting to Pollution," *Scientist and Citizen* 10 (1968):1–8. See also A. V. Colucci et al., "Pollutant Burdens and Biological Response, *Archives of Environmental Health* 27 (1972):151–529.

**Table 4.1**
Symptom Prevalence of Chemically Exposed and Unexposed Reference Populations with a Chemically Exposed Population Receiving Treatment to Reduce Chemical Body Burdens

| Symptom | Chemically Exposed Population | Chemically Unexposed Population | Treatment Group (pre-treatment) | Treatment Group (post-treatment) |
|---|---|---|---|---|
| Rash | 17% | 9% | 18% | 4% |
| Acne | 12 | 5 | 16 | 4 |
| Skin Thickening | 9 | 3 | 9 | 4 |
| Paresthesias (dermal sensations) | 19 | 5 | 14 | 2 |
| Weakness | 13 | 3 | 16 | 4 |
| Uncoordination | 21 | 5 | 7 | 0 |
| Dizziness | 20 | 3 | 18 | 2 |
| Fatigue | 52 | 15 | 79 | 5 |
| Nervousness | 22 | 2 | 14 | 4 |
| Disorientation | 6 | 0 | 11 | 0 |
| Headaches | 41 | 14 | 40 | 9 |
| Joint Pain | 43 | 23 | 5 | 0 |
| Muscle Pain | 23 | 8 | 42 | 5 |
| Abdominal Pain | 13 | 7 | 33 | 11 |
| Constipation | 6 | 2 | 26 | 2 |

Note: There is no statistically significant difference between the symptom prevalence of the chemically exposed population and the pretreatment status of the study population. There is a statistically significant difference ($p < .01$) between the exposed and unexposed populations and between the pre- and posttreatment status of the study population. Each of the four populations had documented fat levels of a variety of foreign chemicals, although the levels in the exposed and the pretreatment groups were significantly higher than the unexposed and posttreatment populations.

Source: D. E. Root, D. B. Katzin, and D. W. Schnare, "Diagnosis and Treatment of patients presenting subclinical signs and symptoms of exposure to chemicals which bioaccumulate in human tissue," *Proceedings of the National Conference on Hazardous Wastes and Environmental Emergencies*, May 14, 1985, Cincinnati, Ohio.

**Table 4.2**
**Percent Reductions in Adipose Tissue Concentrations in PBB-exposed Individuals Receiving a Detoxification Treatment**

| Contaminant | Post-treatment Reduction | Four Month Follow-up Reduction |
|---|---|---|
| Polycholorinated Biphenyls | | |
| 2,3,4,2',4',5'-hexa | 32.8% | 60.6% |
| 2,4,5,2',4',5'-hexa | 17.2 | 27.8 |
| 2,4,5,2',3',6'-hexa | 20.4 | 45.3 |
| 2,3,4,5,2',4',5'-hepta | 34.9 | 29.2 |
| 2,3,4,6,2',3',4'-hepta | 26.2 | 56.5 |
| 2,3,4,5,6,2',5'-hepta | 11.9 | 13.3 |
| 2,3,5,6,3',4',5'-hepta | 37.0 | 59.0 |
| Total PCB (sum of peaks) | 34.2 | 38.4 |
| | | |
| Polybrominated Biphenyls | | |
| 2,4,5,3',4'-penta | 34.0 | 52.1 |
| 2,4,5,2',4',5'-hexa | 25.0 | 65.9 |
| 2,3,4,2',4',5'-hexa | 47.2 | 51.4 |
| 2,4,5,3',4'5'-hexa | +4.2 | 30.3 |
| 2,3,4,5,2',3',4'-hepta | +8.0 | 61.5 |
| 2,3,4,5,2',4',5'-hepta | 36.3 | 37.5 |
| Total PBB (sum of peaks) | 34.5 | 58.7 |
| DDE | 3.5 | 40.2 |
| Heptachlor Epoxide | 31.2 | 37.8 |
| Dieldrin | +3.9 | 10.1 |

Note: Reductions at posttreatment were statistically significant compared with pretreatment levels, and reductions at the four month follow-up point were also significant compared with the posttreatment levels ($p < .01$ for both).

Source: D. W. Schnare et al., "Human Body Burden Reductions of PCBs, PBBs and Chlorinated Pesticides," *Ambio* 13:378–380, 1984.

7. D. B. Tuey and H. B. Matthews, "Distributions and Excretion of 2,2',4.4',5.5'—hexabromobiphenyl in Rats and Man," *Toxicology and Applied Pharmacology* 53 (1980):420–31.

8. See, for example, G. Lambert and J. Brodeur, "Influence of Starvation and Hepatic Microsomal Enzyme Induction of the Mobilization of DDT Residues in Rats," *Toxicology and Applied Pharmacology* 36 (1976):111–20; A. Wirth, G. Schlierf, and G. Schettler, "Physical Activity and Lipid Metabolism," *Klinica Wochenschr* 57 (1979):1105–1201; R. Swartz and F. Sidel, "Effects of Heat and Exercise on the Elimination of Pralidoxime in Man," *Clinical Pharmacology Therapy* 14, no. 1 (1973):83–89; and E. Masoro, *Physiological Chemistry of Lipids in Mammals* (Philadelphia: E. B. Saunders, 1968).

9. U.S. EPA, *Chemicals Identified in Human Biological Media*.

10. L. Golberg, "Safety of Environmental Chemicals—The Need and the Challenge," *Food and Cosmetic Toxicology* 10 (1972):523–29.

11. C. L. Mitchell and H. A. Tilson, "Behavioral Toxicology in Risk Assessment: Problems and Research Needs," *CRC Critical Reviews in Toxicology* 9 (1982):265–74.

12. Golberg, "Safety of Environmental Chemicals."

13. N. K. Mello, "Behavioral Toxicology: A Developing Discipline," *Federation Proceedings* 34, no. 9 (1975):1832–34.

14. B. J. Weiss et al., *Effects on Behavior: Principals for Evaluating Chemicals in the Environment*, (Washington, D.C.: National Academy of Science, 1975).

15. H. C. Scharnweber, G. N. Spears, and S. R. Cowles, "Chronic Methyl Chloride Intoxication in Six Industrial Workers," *Journal of Occupational Medicine* 5 (1963):117.

16. A sample of such studies would include: E. Richter et al., "Paraffin Stimulated Excretion of 2, 4, 6, $2^1$, $4^1$ = Pentachlorobiphenyl by Rats," *Toxicology and Applied Pharmacology* 50 (1979):17–24; K. Rozman, T. Rozman, and H. Grein, "Enhanced Fecal Elimination of Stored Hexachlorobenzene from Rats and Rhesus Monkeys by Hexadecane or Mineral Oil," *Toxicology* 22 (1981):33; and J. Street, "Methods of Removal of Pesticide Residues," *Canadian Medical Association Journal* 100 (1969):16.

17. See: R. Kimbrough et al., "The Effect of Different Diets or Mineral Oil on Liver Pathology and Polybrominated Biphenyl Concentration in Tissues," *Toxicology and Applied Pharmacology* 52 (1980):442.

18. Lucas, *PCBs in Human Adipose Tissue and Mother's Milk*.

19. Y. Oschry and B. Shapiro, "Fat Associated with Adipose Lipase, the Newly Synthesized Fraction that is the Preferred Substrate for Lipolysis," *Biochimica Biophysica Acta* 664 (1981):201–206.

20. For a discussion of mobilization and excretion of xenobiotics stored in human fat, see: D. W. Schnare et al., "Evaluation of a Detoxification Regimen for Fat-Stored Xenobiotics," *Medical Hypotheses* 9 (1982):265–82.

21. R. Warner, "Treatment of PCP Exposure in Narcotics Officers," *Journal of the Fraternal Order of Police*, Winter 1983, p. 18.

22. Schnare, "Evaluation of a Detoxification Regimen."

23. See: D. W. Schnare and P.C. Robinson, "Reduction of the Human Body Burdens of Hexachlorobenzene and Polychlorinated Biphenyls," in *Hexachlorobenzene: Proceedings of an International Symposium*, IARC Scientific Publications No. 77, ed. C. R. Morris and J. R. P. Cabral, (Lyon, France: International Agency for Research on Cancer, 1986), p. 597; D. W. Schnare et al., "Human Body Burden Reductions of PCBs, PBBs, and Chlorinated Pesticides," *Ambio* 13 (1984):378–80; D. E. Root and J. Anderson, "Reducing Toxic Body Burdens Advancing In Innovative Technique," *Occupational Health & Safety* 2, no. 4 (1986):5–8; M. Ben, "Is Detoxification a Solution to Occu-

pational Health Hazards?" *National Safety News*, May 1984; D. E. Root, D. B. Katzin, and D. W. Schnare, "Diagnosis and Treatment of Patients Presenting Subclinical Signs and Symptoms of Exposure to Chemicals which Bioaccumulate in Human Tissue," in *Proceedings of the National Conference on Hazardous Wastes and Environmental Emergencies*, Cincinnati, Ohio, May 14, 1985.

# 5

# The Toxic Tort Is Ill: Deficiencies in the Plaintiff's Case and How to Prove Them

## BARBARA P. BILLAUER, AVRAHAM C. MOSKOWITZ, AND KAREN L. I. GALLINARI

In recent years, the public perception of the chemical using and producing industry has been shaped by such tragedies as Bhopal, Times Beach, Love Canal, and Chernobyl. Along with the adverse publicity generated by such catastrophes, a spate of lawsuits alleging personal injury or property damage has been filed as a result of exposure to chemicals or radiation released into the environment. While many of the lawsuits filed are traditional in nature, in that the plaintiffs have already suffered bodily injury or property damage, some of the lawsuits seek to recover damages for novel "injuries" such as an increased risk of contracting disease, the need for medical monitoring, and "cancerphobia." These new types of claims present unique legal and societal challenges for those involved in toxic tort cases.

The plaintiff's use of epidemiology and risk assessment as support for these claims is as novel as the claims themselves. This chapter argues that the speculative and unreliable nature of this type of evidence makes it an insufficient basis for plaintiffs to prove their similarly speculative claims.

The reader is encouraged to keep in mind the importance of requiring plaintiffs to present reliable, sufficient, and credible evidence to support their claims, and the potential results of failing to do so. If defendants are forced to pay damages to plaintiffs whose only claim is the *possibility* of future injury, or who present insufficient evidence to establish a causal connection between a present injury and defendants' negligence, they will run out of funds to compensate plaintiffs who have viable claims and actual injuries caused by the defendants. Other results include increased prices to the consumer, as well as potential insolvency for the defendant manufacturers.[1]

## EPIDEMIOLOGY AND RISK ASSESSMENT

"Epidemiology [is] a specialized field of medicine dealing with public health based on the observation of the occurrence of disease and thereafter, by statistical methods trying to arrive at a conclusion as to the possible source of the disease."[2] Epidemiology therefore establishes associations between a suspected cause and an identified result, the strength of which is dependent on ruling out other possible causes, the size of the population studied, and the degree to which one can quantify the cause and document the response. Strict proof of "causation" in the legal or logical sense is not the object of the study.[3] As there are always individuals who contract diseases without demonstrable proof of exposure, and as outside confusing variables (confounding factors) cannot usually be ruled out, epidemiologists term the relationship between substance and result as "association" rather than "cause."

Difficulties arise when epidemiological evidence is used to retrospectively assess causation where there are multiple substances known to be associated with, or to increase the risk of, contracting a particular disease. In such situations, epidemiological input can only crudely assign percentages of causation or attributability.[4] Therefore, when epidemiological evidence is admitted at a trial, speculation by the jury becomes unavoidable.

Quantitative risk assessment is the mathematical calculation of human risk that "consists of extrapolation of experimental results achieved at high dose levels to much lower dose levels within the species tested, and then extrapolation of these estimated risks at low doses [in test animals] to [the] risks . . . humans [would face]."[5] Quantitative risk assessment takes epidemiology one step further in an effort to predict the *number* of people in a given population who *may* contract a particular disease, but stops short of naming which individuals *will* get ill.

A report issued by the federal government's Interagency Regulatory Liaison Group, which discusses the methods used in estimating the carcinogenic potential of a substance, indicates that quantitative risk assessment looks at the worst case and calculates a risk factor for a hypothetical population.[6] It does not consider risks to specific individuals.

As Chapter 3 makes clear, mathematical modeling used in epidemiology and risk assessment is unable to account for individual variables such as physical differences, repair mechanisms, genetic susceptibility, immune functions, and exposure to predisposing substances or localities. While it may not be necessary to account for these differences in a person who already has a particular disease, these variables do have great impact on predicting whether a specific person, or group of people, will contract a disease in the future.

In addition to the problems mentioned above, both epidemiological and risk-assessment analyses rest upon numerous assumptions including the mechanism of how cancer is caused. Assumptions, of course, decrease the reliability of the results. Since "there is no unanimity of opinion as to the method of choice for cancer risk assessment," the choice of methodology may mean great variation in predicting results or in establishing the strength of association or causation.[7]

As Cothern and Schnare make clear in Chapter 3, these methodologies may be useful in the development of public health guidelines, but they cannot be relied upon as a means of assessing actual causation in a particular individual with the degree of certainty required by law. These studies are incapable of predicting whether a particular member of a population will contract a disease in the future.

As discussed above, both epidemiology and quantitative risk assessment deal only with mathematical predictions of disease in the general populations and depend upon numerous assumptions. They are, therefore, inherently unfit as proof that exposure to a particular substance caused a disease in a particular person. They are even more unfit to prove that a person *will* contract a disease in the future due to exposure to a particular substance. Since each plaintiff must prove by a "fair preponderance of the credible evidence" that it is "reasonably certain" or "reasonably probable" that he has or will sustain quantifiable damage, we argue that epidemiological studies and quantitative risk assessments must be found insufficient, as a matter of law, to sustain plaintiff's burden of proof. Nevertheless, plaintiffs are still forced to use such evidence, for without the use of these studies, the plaintiff's claims for increased risk of fear of future disease would be completely unsupported and necessarily would fail.

## EVIDENTIARY PRINCIPLES

Claims for exposure to toxic substances must meet traditional prerequisites of civil litigation. In order to prevail in toxic tort cases plaintiffs must establish at least two legal requirements: causation and damages.[8] They must prove these elements by a fair preponderance of the credible evidence, meaning "more probably than not" they have sustained measurable damages proximately caused by the defendant. To do this plaintiffs must produce legally admissible forms of evidence. The Federal Rules of Evidence outline the prerequisites for admissible evidence.[9]

To be recognized as admissible, the evidence (testimony or documentation) must be relevant and reliable, and must also not mislead or unfairly prejudice the jury.[10] Therefore, courts are averse to admitting evidence that is highly speculative.[11] Courts do not want the jury to be misled by information when its accuracy is in question, since the jury's reliance on such evidence will lead to speculative verdicts.[12]

Thus, before evidence will be admitted, the court "must assess the admissibility of [any] testimony based on a novel scientific technique by balancing the relevance, reliability, and helpfulness of the evidence against the likelihood of waste of time, confusion and prejudice."[13]

### Reliability

To assure the reliability of novel scientific evidence the majority of courts apply the ruling first articulated in *Frye v. United States*:[14] "While courts will go a long way in admitting expert testimony deduced from a *well-recognized scientific principle or discovery*, the thing from which the deduction is made must be sufficiently established to have *gained general acceptance* in the particular field in which it belongs."[15]

In a recent proceeding in the Agent Orange litigation, Chief Judge Weinstein pointed out the problems that make many epidemiologic studies legally inadmissible to prove causation.[16] The plaintiffs attempted to rely on many studies that "involved laboratory animals subjected to extreme exposures [to Agent Orange] with unknown human significance; some, . . . [which] have never been replicated, and involved chemicals in addition to the constituents of Agent Orange . . . [and] others [which] involved chronic or acute industrial exposures different from the exposures in Vietnam."[17]

Chief Judge Weinstein found that the plaintiff failed to establish a material issue of fact that could be submitted to a jury. Interestingly, the court based its conclusion not only on the poor quality of the plaintiff's evidence, but also on the impressiveness of the epidemiological evidence submitted by the defendants. The defendants' epidemiological evidence consisted of a group of studies that had been conducted by several different agencies, including state and federal agencies of the U. S. government, as well as an agency of the Australian government. The court noted that these studies did not suffer from the infirmities evident in the plaintiff's studies since, for example, they studied the effects of Agent Orange on Vietnam veterans. These government studies, which were found to be relevant and reliable, demonstrated that based on present knowledge, Agent Orange cannot be proved to have caused plaintiff's illness.

In addition to noting the quality of the epidemiological reports submitted by the defendants, Chief Judge Weinstein applied the rule of evidence which provides that where an epidemiological study is done by a public agency, there is a presumption of admissibility countered only by showing the report's untrustworthiness.[18] Thus, the source of the epidemiological evidence lends some presumptive reliability to it.

This rule was also applied in *Kehm v. Procter & Gamble Manufacturing Co.*, in which defendants argued that plaintiffs' epidemiological evidence was unreliable and hence inadmissible because it rested on medical opinions and diagnosis rather than on facts.

The court there held that since the findings "are public records based on investigations conducted pursuant to lawful authority, they are presumptively reliable. That is, 'there is no reason not to admit the findings simply because they tend toward the conclusory rather than the factual end, unless . . . the sources of information or other circumstances indicate lack of trustworthiness.' "[19]

## Sufficiency

In addition to the requirements that the plaintiff's evidence be relevant and reliable, it is also necessary that the evidence be *sufficient* to prove that the defendant's conduct was *more likely than not* the cause of the plaintiff's harm.[20] The courts do not want the jury to speculate upon any evidence that does not meet this standard, and judges frequently articulate its application.

"The causal relationship of an accident or injury to a resulting physical condition must be established by medical testimony beyond speculation and conjecture.... It must rise to the degree of proof that the resulting condition was probably caused by the accident."[21] In *Mullaney v. Goldman*, the court held that "in those cases where expert testimony is relied on to show that out of several potential causes a given result came from one specific cause the expert must report that the result in question '*most probably*' came from the cause alleged."[22]

In his opinion approving the Agent Orange settlement, Chief Judge Weinstein stated:

The evidence provided by the plaintiffs to date on general causality, while supportive of the desirability of further studies, lacks sufficient probative force... to permit a finding of general causality. It might require the direction of a verdict for defendants at the end of plaintiffs' case. It simply is not sufficient to point to an individual and show that he was exposed to Agent Orange and had a cancer. The incidence of cancers of the type suffered by plaintiffs in the population as a whole make it at least as likely, based upon present knowledge, that the cancer resulted from causes other than Agent Orange.[23]

The insufficiency of risk-assessment evidence was recognized in *Herber v. Johns-Manville*.[24] The court, in barring plaintiff from introducing evidence at trial relating to an increased risk of contracting cancer, stated:

Assuming, as plaintiff conceded, that he has an increased *possibility* of developing cancer, it is clear that as a matter of law this possibility is an insufficient foundation upon which to base an award of damages because such damages would be based on mere conjecture, impermissible under New Jersey law. That he has a much greater risk of developing cancer than a 46-year-old white male who has not been exposed to asbestos and/or smoked, does not make the foundation more sufficient for the test is reasonable probability as to him.[25]

Similarly, the court in *Devlin v. Johns-Manville Corp.* held that testimony of plaintiffs' expert that they had an increased *possibility* of contracting cancer was insufficient to support a cause of action.[26]

In most toxic cases, plaintiff and his experts must rely exclusively on epidemiological studies and quantitative risk assessments to attempt to prove that plaintiff is likely to contract a disease as a result of his exposure to a particular toxic element because it is the only evidence available to the plaintiff.

Due to its inherent limitations, the use of epidemiological evidence is problematic along every step of this legal obstacle course. It can be argued that (1) it is inherently *unreliable* because of lack of uniformity in use in the scientific community; and (2) it cannot predict individual responses rendering it *insufficient* and speculative.[27]

## EPIDEMIOLOGICAL EVIDENCE IN THE COURTROOM

Plaintiffs have attempted to use epidemiological evidence and risk-assessment analyses to prove causation where injury has already occurred and to establish a likelihood of additional future injury in those cases. It is common for these claims for present and future injury to be presented to the court in the same lawsuit. For purposes of deciding whether the claim is compensable and whether the plaintiff's evidence will be admitted, the courts generally deal with these claims individually.

### Claims of Present Injury

A plaintiff who currently suffers from a physical illness or from mental distress allegedly caused by exposure to a toxic substance must establish a causal relationship between the toxic substance and the injury in order to have a viable claim for present injury. A plaintiff with a toxic exposure claim is likely to present epidemiological evidence to prove causation between the exposure to the substance and his injury.

Where epidemiological data is used in an attempt to establish a causal relationship some courts have ruled that such data is inadmissible because it is too speculative, and could not adequately rule out the possibility that other causes are responsible for plaintiff's disease.

*Claims for Present Physical Disease.* In *Parker v. Employers Mutual Liability Ins. Co. of Wisconsin*, the court reversed a plaintiff's verdict because plaintiff's medical experts testified that plaintiff's cancer *could* have been caused by his exposure to radiation in the workplace.[28] Plaintiff's experts did not testify that his cancer was *probably* caused by such exposure. In its decision the court noted there can be many possible 'causes,' indeed, an infinite number of circumstances can cause an injury. But a possible cause only becomes 'probable' when in the absence of other reasonable causal explanations it becomes more likely than not that the injury was a result of its action. This is the outer limit of inference upon which an issue can be submitted to the jury."[29]

Similarly, in *Gardner v. Hecla Mining Co.*, the court upheld a denial of workers' compensation benefits to the plaintiff who claimed that his lung cancer was caused by radiation exposure during the course of his employment as a uranium miner.[30] The plaintiff had introduced statistical data and expert testimony that uranium miners had a much higher incidence of lung cancer than the general population. After noting that plaintiff had smoked one pack of cigarettes every day for twenty years, the court held that the plaintiff had failed to rule out other possible causes of his cancer and thus had failed to establish that his illness was actually caused by his radiation exposure.[31]

*Claims for Present Mental Distress.* Along with claims of present injury plaintiffs bring claims for mental distress allegedly caused by exposure to toxic substances. The plaintiff claims that due to his exposure to an allegedly carcinogenic substance he presently suffers from the fear that he will contract cancer in the future. While often interchangeably referred to as "cancerphobia" or "fear of cancer" claims, there is an important distinction between these two conditions that impacts on the plaintiff's level of evidentiary proof.

"Cancerphobia" is a psychiatric illness characterized by an exaggerated and irrational response to the knowledge that one is at risk of contracting cancer, and requires expert testimony usually by a psychiatrist to substantiate it.[32] (For example, the plaintiff might claim that he no longer leaves the house for fear of being exposed to one more toxic substance.)

"Fear of cancer" is considered a rational response to the same knowledge. "A plaintiff [bringing a claim for fear of cancer] can testify to his fear, preoccupation and distress resulting from the enhanced risk of cancer about which he has been informed."[33] Since this is classified as a common experience that lay people are competent to evaluate, no expert testimony is required.[34]

The claims for nervous disorders are often superimposed upon claims for a present bodily injury by reason of legal necessity. While the law regarding recovery for mental distress is currently in flux, to safeguard against fraudulent claims, the majority of courts will not permit recovery for mental distress "without accompanying physical injury, illness or other physical consequences."[35] Many courts additionally require evidence of a physical " 'impact' upon the person of the plaintiff."[36] The great majority of courts however limit themselves to the requirement of physical injury, illness, or other physical consequence.[37]

According to the United States Court of Appeals for the Fifth Circuit, "recent commentators have noted, 'courts having considered cancer phobia [the present anxiety over developing cancer in the future'], almost uniformly have allowed it" if an underlying physical condition is present and if the plaintiff can sustain the necessary quantum of evidentiary proof needed to show the plaintiff's fear is rational, based on the likelihood he will contract future disease.[38] For example, in *Gideon v. Johns-Manville Sales Corp.*, the court found that Texas law permits recovery for fear of future conditions that will in medical *probability* develop from presently existing injuries, namely asbestosis.[39]

In *Dartez v. Fibreboard Corp.*, however, the court permitted the plaintiff to present epidemiological evidence of increased risk, through expert testimony, to support a claim for mental anguish, even though that evidence failed to prove that he would *probably* contract the disease of which he was afraid. The court stated:

Mental anguish would reasonably follow after Dartez was informed by a reliable source—his physician—that his exposure to defendants' products had heightened his risk of developing these deadly diseases. Under Texas law he is entitled to compensation for mental anguish proximately caused by his asbestos exposure, even if such distress arises from fear of diseases that are a substantial concern, but not medically probable. The fact that the evidence does not establish a reasonable medical probability that Dartez will develop either disease is relevant to the jury's determination of the reasonableness of the anguish he feels and the amount of compensation to which he is entitled. However, it does not suffice to bar the proof for purposes of establishing mental anguish.

Dartez's own testimony regarding his emotional reaction to the realization of the risks he now faces and Comstock's testimony as to what he told Dartez about these risks were properly admitted to establish the mental anguish element of plaintiff's case.[40]

Thus, in *Dartez* the court assured itself that the plaintiff's physical claim was real and then treated the plaintiff's mental claim as a present and rational condition based on what was told to him by his doctor. The issue of probability of future harm was treated as an index of the depth and severity of plaintiff's present fear—rather than as an index of whether he was likely to contract a disease in the future. Epidemiological evidence has been used as a method of proving the rationality of plaintiff's claim of present fear to show a connection between a present claim and the present fear of the possibility (as opposed to probability) of developing future injury.

### Claims of Increased Risk of Future Injury

*Increased Risk Claims.* In addition to using epidemiology to sustain a claim for *present* fear based on a statistical connection between a present physical ailment and the possibility of future physical ailment, an even more novel concept using epidemiological evidence has been espoused by plaintiffs. Many plaintiffs now try to say that solely by virtue of exposure to a toxic chemical they are in a special category of "risk of disease" which creates a new claim. This concept involves a more speculative use of epidemiological evidence than using epidemiology to establish likely retrospective association (causation) or to validate a present fear of contracting a future disease. Since such plaintiffs are not presently suffering from that disease, they must use epidemiological studies and risk-assessment analyses to predict a future likelihood. Not only has the question of the reliability of the evidence become critical to this determination, but the sufficiency of whether epidemiology can predict this information by a reasonable probability must be carefully addressed before such evidence can be admitted.

Where a plaintiff establishes that he suffers from a present physical illness caused by exposure to the toxic agent, many courts will allow his claim for increased risk of future disease if the plaintiff can establish that the alleged future harm is "likely to" or "probably will" occur. Absent a present injury most courts reject claims for future injury.[41]

Once some physical effect of a toxic exposure appears, "[t]he general rule is that where it is established that future consequences from an injury to a person will ensue, recovery therefore may be had, but such future consequences must be established in terms of reasonable probabilities."[42]

Most courts require that in order to meet this "reasonable probability" standard the plaintiff's epidemiological evidence must prove that the plaintiff has a greater than 50 percent chance of contracting cancer in the future (that is, the evidence proves that the plaintiff will "more likely than not" contract cancer in the future).[43]

In *Gideon* the plaintiff presented evidence that he currently suffered from asbestosis, *and* that he had a greater than 50 percent chance of contracting an asbestos-related cancer. The court stated that, "in Texas, as elsewhere, the time honored method of proving future physical condition is to present a qualified physician's opinion testimony based on reasonable medical probability. Possibility alone cannot serve as the basis for recovery, for mere possibility does not meet the preponderance of the evidence standard. Certainty, however, is not required: the plaintiff need demonstrate only that the event is more likely to occur than not."[44]

In *Arnett v. Dow Chemical Co.*, the court expressed severe reservations about the use of quantitative risk assessment to prove that the plaintiff had an increased risk of contracting cancer. The *Arnett* court stated that

the developments at the very frontier of science do not provide reasonably probable predictions. There is no definitive epidemiological evidence to verify the mathematical calculation of quantitative risk assessment. While the animal test results and *in vitro* studies present the possibility that DBCP *may* be a human carcinogen, such an extrapolation does not reach the requisite level of acceptance within the scientific community to justify legal reliance.[45]

Further, at least one court has decided that even if the plaintiff presents evidence that he has a greater than 50 percent chance of contracting cancer, a claim for increased risk of future disease should *not* be allowed.

The speculative nature of the prediction of future damages . . . may lead to several inequitable results. First, the plaintiff who does not contract cancer gets a windfall—cancer damages without cancer. Second, and perhaps worse, an asbestosis plaintiff who is unsuccessful in his efforts to recover risk of cancer damages, but later contracts cancer, has the disease, but no damages. Third, even plaintiffs who later contract cancer and who have recovered some amount of risk of cancer damages may emerge with an inequitable award, since the jury, cognizant of the less than one hundred percent chance that the plaintiff will contract cancer, likely will have awarded less than one hundred percent damages. Finally, inequitable awards are more likely to result from a future damages action simply because the damages cannot be known.[46]

*Claims for Medical Surveillance.* In some cases, plaintiffs who claim that they are at an increased risk of developing cancer try to recover for the costs of ongoing medical monitoring that, allegedly, their doctors have recommended. To date, some courts have recognized that recovery for the costs of medical surveillance is permissible, but recovery is limited by the speculativeness of the plaintiff's claims.

For example, in the toxic waste case, *Ayers v. Township of Jackson*, the lower court found that public policy dictates recovery for medical surveillance when in reasonable medical judgment it is found necessary.[47] The appellate division vacated this award stating "without some quantifying guidance it becomes impossible to say that defendant has significantly increased the 'reasonable probability,' ... that any of the plaintiffs will develop cancer so as to justify imposing upon the defendant the financial burden of lifetime medical surveillance for early clinical signs of cancer."[48]

Similarly, in *Friends For All Children v. Lockheed Aircraft*, the court found that although both the tort law of the District of Columbia and general tort principles permit recovery for diagnostic examinations even in the absence of actual physical injury, recovery could only be allowed when some injury to the plaintiff is "neither speculative nor resistant to proof."[49] Nevertheless, the court did permit the costs of a one-time diagnostic examination to be awarded to Vietnamese orphans who possibly sustained injuries during depressurization when the plane they were traveling in crashed. It would seem this award was based more on the "interests of justice" and compassion than on any recognized legal theory.

Other courts have permitted plaintiffs to recover for the costs of medical surveillance but have insisted that recovery be allowed only when there is clear evidence that the medical surveillance is necessary based on reasonable medical judgement.[50] The plaintiff must present a medical expert who has recommended medical surveillance, based upon *acceptable* scientific evidence, before recovery will be permitted. If the plaintiff's expert is relying upon epidemiologic studies and risk assessment, indicating an increased risk of disease, which are fraught with too much uncertainty and speculation, the court should deny recovery.

## INCREASED RISK CLAIMS WITHOUT PRESENT INJURY

The law of torts has traditionally required that the plaintiff prove, by a fair preponderance of credible evidence, that he has suffered an *actual injury* before damages could be recovered. In the words of Dean Prosser, "the threat of future harm, not yet realized, is not enough."[51] Despite this general rule, persons who have not suffered any personal injury but who have allegedly been exposed to chemicals or other pollutants released into the environment have begun asserting claims for an alleged increased risk of contracting disease even where no present actual injury exists. Plaintiffs use epidemiological evidence and risk assessment as a basis for these claims. They try to assert that past epidemiological studies show an increased incidence of disease in those exposed and consequently predict—with varying levels of probability—whether someone exposed will contract the disease in the future (that is, in that person's lifetime).

Although this new type of claim has become increasingly popular in various contexts, it has been rejected by the courts in many jurisdictions. For example, in *Laswell v. Brown*, the court rejected plaintiffs' claim for damages based on their having an increased risk of contracting cancer as a result of their exposure to radiation during the U.S. Army's testing of nuclear weapons. The court held that "[t]he 'complaint is conspicuously void of any allegations that the children have sustained any damage other than the exposure to a higher *risk* of disease and cellular damage.' We agree with the district court that a lawsuit for personal injuries cannot be based only upon the mere possibility of some future harm."[52]

Similarly, in *Ayers v. Township of Jackson*, 325 residents of Jackson Township filed suit against the defendant township alleging that toxic wastes leached through a municipal landfill, owned and operated by the defendant, and contaminated the plaintiffs' well water, thereby allegedly subjecting them to an increased risk of contracting cancer.[53]

The defendants moved for summary judgement with respect to the claim for increased risk of cancer on the grounds that such a claim was not recognized under the law of New Jersey or under federal law.

In reviewing the evidence the plaintiffs intended to introduce at trial, the court noted that plaintiffs' expert witnesses would testify "that all individuals exposed to the well water contamination are at an increased risk of developing cancer and liver and kidney damage."[54] Based on this testimony, plaintiffs argued that their increased risk of contracting cancer was a presently *existing* condition that constituted a compensable injury.

The court granted the defendant's motion for summary judgement and rejected plaintiffs' claim based on an increased risk of cancer, stating that

plaintiffs characterize the increase in the risk of suffering a disease as reasonably probable and, therefore, compensable. But that risk cannot be quantified nor do any of the plaintiffs' experts even attempt to advance an opinion that any of the 325 plaintiffs have or will probably contract any of the diseases. The trier of fact is left to speculate as to the possible consequences of ingestion of the alleged carcinogens and other chemicals to the future health of each plaintiff. . . .

To permit recovery for possible risk of injury or sickness raises the spectre of potential claims arising out of tortious conduct increasing in boundless proportion. Without minimizing plaintiffs' claim, the court cannot ignore the fact that much of what we do and make part of our daily diet exposes us to potential, albeit remote, harm. As long as the risk exposure remains within the realm of speculation, it cannot be the basis of a claim of injury against the creator of that harm.[55]

On appeal, the Appellate Division of New Jersey's Superior Court upheld this reasoning stating: "As we explained in connection with plaintiffs' claims for future medical surveillance, the degree of increased risk was in no way quantified. Indeed, that function was described by plaintiffs' expert witness as 'impossible,' and we therefore conclude that a reasonable probability of enhanced risk is not supported by the evidence. We discern no way to compensate one for enhanced risk without knowing in some way the degree of enhancement."[56]

Similarly, in *Arnett v. Dow Chemical Co.*, the court stated: "to award damages based on a mere mathematical probability [of future injury] would significantly undercompensate those who actually do develop cancer and would be a windfall to those who do not."[57]

Similar rulings have also been rendered in cases where plaintiffs have sought recovery for an increased risk of contracting cancer as a result of ingesting or being exposed, *in utero*, to the drug DES. In *Plummer v. Abbott Laboratories*, plaintiffs did not allege that they had suffered physical injury but rather alleged a heightened risk of contracting cancer. The court dismissed the plaintiffs' claim, reasoning that "the possibility that an individual may, because of the ingestion of certain drugs, have acquired a greater risk of contracting cancer does not *per se* constitute injury for purposes of tort law."[58] In *Mink v. University of Chicago*, the closest plaintiffs came to "alleging physical injury" was the claim for an increased " 'risk' of cancer," thus leaving the court to conclude that "the mere fact of risk without any accompanying physical injury is insufficient to state a claim."[59] Similarly, in *Morrissy v. Eli Lilly & Co.*, the court stated that "the nexus ... suggested between exposure to DES *in utero* and the possibility of developing cancer or other injurious conditions in the future is an insufficient basis upon which to recognize a present injury."[60] Likewise, in *Rheingold v. E. R. Squibbs & Sons, Inc.*, the court granted defendants' motion to dismiss because plaintiff had not pleaded injury to anyone but only claimed a future risk of contracting cancer.[61]

These cases indicate that, absent present injury to plaintiff, the courts in many jurisdictions will look askance at a claim based purely on an increased risk of future injury.[62]

A variation of the claim for increased risk of contracting disease in the future is currently being litigated in a property damage context. An example of this type of case is the litigation between school districts, towns, and municipalities, and manufacturers of asbestos-containing products.[63] In these cases, the plaintiffs rely on the most speculative use of epidemiological evidence to predict the number of people who will contract a disease based on possible exposures many hundreds of times less than current epidemiological data provides. The plaintiffs are seeking to recover the cost of removing or repairing products containing asbestos that were installed in schools and other public buildings on the basis that exposure to ultra-low levels of asbestos increases one's risk of contracting an asbestos-related cancer. Plaintiffs argue that these asbestos-containing products must be removed from plaintiffs' buildings to prevent such a possibility.

By 1988, only four asbestos property damage cases have gone to verdict, resulting in two verdicts in favor of the defendants.[64]

In the *County of Anderson, Tennessee* case, the defendants countered the plaintiff's arguments by introducing evidence comparing the relative risk of contracting cancer associated with a number of other commonplace exposures.[65] Thus, for example, the risk of contracting lung cancer from smoking a pack of cigarettes a day for twenty years is 88,000 per 1 million persons.[66] In contrast, the risk of contracting cancer from eating peanut butter containing aflatoxin at the legally permissible level, is 11 per 1 million persons, while the risk associated with exposure to asbestos at the levels found in the plaintiffs' school buildings is 0.8 per 1 million persons.[67] This evidence may have helped convince the *Anderson* jury to put the concept of risk assessment in proper perspective.

Recently, two courts have dismissed complaints alleging asbestos-related property damage claims.[68] In one of these cases, Judge Curry of the Circuit Court of Cook County, Illinois, compared the plaintiffs' complaints to the shelves in a little dark shop that perplexed Alice in Wonderland. "Whenever you look hard at any count to make out exactly what it had in it, that particular [count] was always 'quite empty.' " Judge Curry concluded his opinion by criticizing the plaintiffs for

invit[ing] the court to enter the twilight zone of legal remedies where damages are defined by threat or risk rather than by harm or injury. That novel deviation would then, I suspect, give way to statistical probabilities of damage, which in turn would be replaced by a presumption of damage and so on until we arrived at the Shangrila of guaranteed awards.
. . .
Plaintiffs have shouted "asbestos" in a crowded schoolhouse in an attempt to panic the courts into adopting hybrid procedures, carving out special categories, stretching known legal precepts beyond logic or necessity and abandoning antiquity, uniformity and universality in the law.[69]

## CLAIM SPLITTING AS A VIABLE SOLUTION

Plaintiffs contend that the refusal to accept epidemiological evidence and causes of action for future harm leaves plaintiffs without a remedy if indeed future harm develops. Some courts have refused to exclude evidence that the plaintiff is at a higher risk of contracting cancer when a plaintiff has some present injury due to a toxic exposure on the basis that a plaintiff should not be permitted to come to court a second time to argue his case twice. This is based on the traditional principal of tort law that required a plaintiff to bring all his claims against a defendant who harmed him in one lawsuit. "Once injury results there is but a single tort and not a series of separate torts, one for each resultant harm."[70] Despite the fact that cancer is not yet a "resultant harm," many courts are bound to follow the laws in certain jurisdictions that consider reasonably certain future injuries as part of the plaintiff's single cause of action.

This concept may explain the ruling of the court in *Gideon v. Johns-Manville Sales Corp.*[71] The court held that the law of Texas prevented the plaintiff from suing the defendant a second time (after suing for asbestosis) if he later developed cancer from his original asbestos exposure. The court found that:

Under Texas law, ... Gideon has but one cause of action for all the damages caused by the defendants' legal wrong, the diseases that have developed and will in probability develop are included within this cause of action, for they are but part of the sequence of harms resulting from the alleged breach of legal duty [citation omitted]. Gideon could not split his cause of action and recover damages for asbestosis, then later sue for damages caused by such other pulmonary disease as might develop. Then still later sue for cancer should cancer appear.[72]

Other courts have excluded evidence of increased risk of cancer based upon a new development in recent case law, which gives the plaintiff a new cause of action upon diagnosis of cancer.[73] This alternative, commonly referred to as "claim splitting," may prove to be beneficial from a defense standpoint.

For example, in *Eagle-Picher Industries, Inc. v. Cox*, the District Court of Appeals of Florida decided that the plaintiff could not recover damages for his alleged increased risk of contracting cancer in the future, but if he should later contract cancer, the rule against splitting causes of action would not bar the plaintiff from bringing a second action seeking damages for cancer.[74] The court found that blindly adhering to the rule against claim splitting in cases of alleged increased risk of disease only served to encourage the use of speculative testimony and would likely lead to inequitable results. The court stated: "We have come to our decision that any recovery for cancer damages must await the actuality of cancer by balancing the need for finality against countervailing factors which militate in favor of the splitting of the actions. This balancing process recognizes that the desirable goal of [legal] finality is not an absolute and that the procedural rule against splitting causes of action must be replaced when equitable considerations demand it."[75]

The court found ample authority for its decision to reserve that plaintiff's right to maintain an action should cancer actually occur. Most notably it pointed out that according to current law the general rule against action splitting is inapplicable where

—The court in the first action has expressly reserved the plaintiff's right to maintain the second action; or

—It is clearly and convincingly shown that the policies favoring preclusion of a second action are overcome for an extraordinary reason.[76]

The court found these exceptions particularly appropriate in cases where the reservation of the right to sue for cancer, if and when it occurs, helps to justify the preclusion of the right to sue for inchoate cancer claims.

The decisions favoring claim splitting, at first, seem to broaden the recovery allowed to the plaintiff by allowing him to bypass a stringent statute of limitations rule and by giving the plaintiffs "two bites at the apple." Counterbalancing this is the fact that exclusion of evidence of the plaintiff's increased risk of cancer at the original trial could lead to lower verdicts and settlements and ultimately avoid recovery where plaintiffs never become ill. As the *Cox* court noted it will "prevent rather than promote vexatious litigation and a multiplicity of lawsuits."[77]

## CONCLUSION

In most cases in which plaintiffs allege an increased risk of disease as a result of exposure to allegedly toxic chemicals, a careful analysis of plaintiff's medical evidence may reveal that such evidence is inadmissible at trial because it is not based on methods and theories that have attained general acceptance in the scientific community. Furthermore, even if such evidence is considered scientifically reliable, it may prove to be deficient because it is incapable of establishing that it is "reasonably certain" or "reasonably probable" that plaintiff will contract a disease and that if he does, that such disease was caused by exposure to the allegedly toxic substance in question. Preventing recovery in cases where the plaintiffs claim is tenuous, fraudulent, or trivial will help ensure that there is enough money to compensate those who are seriously ill and whose illness was caused by defendant's negligence as opposed to those who may become ill by their own actions or by acts of God.

## NOTES

The opinions expressed herein are not necessarily those of the office of the United States Attorney for the Southern District of New York nor those of the United States government.

While citations and references to case law and legislation were current at publication, the reader should keep in mind that the law is dynamic and will change over time.

1. See, for example, *In re Johns-Manville Corp.*, nos. 82-B-116 56 through 82-B-11676 (BLR) (Bankr. S.E.N.Y. filed Aug. 26, 1982); *in re UNR Industries, Inc.*, nos. 82-B-9841 through 83-B-9851 (Bankr. N.D. Ill. filed July 29, 1982).

2. *Reyes v. Wyeth Laboratories*, 498 F.2d 1264, 1271 n.3 (5th Cir.), cert. denied, 419 U.S. 1096 (1974).

3. Causation is often defined by a 'but for" test. In the words of Dean Prosser: "The defendant's conduct is a cause of the event if the event would not have occurred but for that conduct; conversely, the defendant's conduct is not a cause of the event, if the event would have occurred without it." W. Prosser and W. Keeton, *The Law of Torts*, 5th ed. (St. Paul, Minn.: West Publishing 1984), sec. 41, p. 266.

4. See P. Enterline, "Asbestos and Lung Cancer: Attributability in the Face of Uncertainty," *Chest* 78 (Supp. 2) (1980):377.

5. Schottenfeld and Haas, "Carcinogens in the Workplace," *CA—A Cancer Journal for Clinicians* 29 (1979):144, 152 (footnotes omitted).

6. *Scientific Bases for Identification of Potential Carcinogens and Estimation of Risks*, 44 Federal Register (1979) p. 39, 858.

7. M. D. Hogan and D. G. Hoel, "Extrapolation to Man," in *Principles and Methods of Toxicology*, ed. A. Hayes (New York: Raven Press, 1982), p. 717.

8. These requirements are common to both negligence and strict product liability claims. See Prosser and Keeton, *Torts*, sec. 30 and sec. 103, respectively.

9. The Federal Rules of Evidence were "enacted into law by Public Law 93-595, 88 Stat 1926, approved January 2, 1975, and effective July 1, 1975." Fed. Rules Evid. Serv., Introductory Statement by the Editors (1982).

10. See Fed. Rule Evid., *Definition of "relevant evidence,"* p. 401; *Testimony by Experts*, p. 702; *Hearsay Exceptions*, p. 803 (18); *Availability of Declarant Immaterial*: Learned Treatises; *Exclusion of relevant evidence on grounds of prejudice, confusion, or waste of time*, p. 403; and *Frye v. United States*, F. 1013 (D.C. Cir. 1923), p. 293.

11. See, for example, *Nunez v. Wilson*, 211 Kan. 443, 507 P.2d 329 (1973); *Fort Worth & Denver Railway Co. v. Janski*, 223 F.2d 704, 707 (5th Cir. 1955); 22 Am. Jur. 2d *Damages* Sections 22-27 (1965 and Supp. 1985).

12. See, for example, *Bigelow v. RKO Radio Pictures, Inc.*, 327 U.S. 251, 66 S. Ct. 574 (1946); *Harmsen v. Smith*, 693 F.2d 932 (9th Cir. 1982), *cert denied*, 104 S. Ct. 89 (1983); *Hawthorne Industries, Inc. v. Balfour Maclaine International, Ltd.*, 676 F.2d 1285 (11th Cir. 1982).

13. *In re Agent Orange Product Liability Litigation, Lilley v. Dow Chemical Co.*, MDL No. 381, 611 F. Supp. 1267, 1279 (E.D.N.Y. 1985), *aff'd*, 818 F.2d 145 (2nd Cir., 1987).

14. *Frye v. United States*, 293 F. 1013 (D.C. Cir. 1923). In this landmark case, the court ruled that the results of a deception test, which was the forerunner of the present-day polygraph, were inadmissible.

15. *Frye v. United States*, 1014. As previously noted, quantitative risk assessment has not gained *general acceptance* in the scientific community. As researchers Hogan and Hoel stated, "there is no unanimity of opinion as to the method of choice for cancer risk assessment." See note 7.

(While Billauer et al. put forward this argument, things may not be quite this simple and clear-cut. The entire regulatory effort of the health regulating federal agencies such as EPA and FDA rely extensively on quantitative risk assessment, and a consensus on appropriate uses and standardized assumptions does exist, as shown in the materials referenced in note 6 above. For additional information on this subject, see Chapter 3 of this book, a presentation on the use of quantitative risk assessment by Cothern and Schnare—Ed.)

16. *Lilley v. Dow Chemical Co.*

17. Ibid, p. 1273.

18. Fed. R. Evid. 803(8)(C). *Hearsay exceptions: availability of declarant immaterial*, Public records and reports.

19. *Kehm v. Procter & Gamble Manufacturing Co.*, 724 F.2d 613 (8th Cir. 1983), and p. 618 (quoting *United States v. American Telephone & Telegraph Co.*, 498 F. Supp. 353, 360 [D.D.C. 1980]); (other citations omitted). See also *Lilley v. Dow Chemical Co.*, MDL No. 381,611 F. Supp. 1267 (E.D.N.Y. 1985), *appeal docketed*, No. 85-6269 (2nd Cir. July 30, 1985), *appeal argued*, (2nd Cir. Apr. 9-10, 1986).

The *Kehm* court rejected the arguments that defendants were unable to cross-examine the sources interviewed for the studies, and that the studies had numerous other statistical biases that rendered them untrustworthy. The court claimed that there were other indicia of reliability that rendered the documents admissible. It noted that the trial court had found that the epidemiological studies employed procedures and methods widely accepted in the field of epidemiology. Defendants conceded this finding and also admitted that epidemiologists regularly rely on studies of this kind. Finally, the court held that the defendants had ample opportunity to challenge the methodology and findings of the government studies through presentation of its own evidence.

20. See *Nieves v. City of New York*, 92 A.D.2d 938, 458 N.Y.S.2d 548 (1983); *Matott v. Ward*, 48 N.Y.2d 455, 423 N.Y.S.2d 645, 399 N.E.2d 532 (1979); see also Committee on Pattern Jury Instructions, Association of Supreme Court Justices, New York Pattern Jury Instructions, Civil Sec. 1:90 (1968 and Supp. 1986).

21. *Carlos v. Cain*, 4 Wash. App. 475, 481 P.2d 945, 947 (Wash. Ct. App. 1971) (quoting a previous ruling in *Miller v. Staton*, 58 Wash.2d 879, 886 365 P.2d 333 [1961]).

22. *Mullaney v. Goldman*, 121 R.I. 358, 398 A.2d 1133, 1136 (1979) (quoting a previous ruling in *Sweet v. Hemingway Transport, Inc.*, 114 R.I. 348, 355, 333 A.2d 411, 415 [1975] [emphasis in original]); see also *Koenig v. Heber*, 84 S.D. 558, 174 N.W.2d 218 (1970); *Kujawa v. Baltimore Transit Co.*, 224 Md. 195, 167 A.2d 96 (1961); *Johnesee v. Stop & Shop Co.*, 174 N.J. Super. 426, 416 A.2d 956 (App. Div. 1980).

23. *Lilley v. Dow Chemical Co.*

24. *Herber v. Johns-Manville*, Civ. No. 80–2081 (D.N.J. Nov. 30, 1984), *vacated and remanded*, 785 F.2d 79 (3d Cir. 1986).

25. Ibid., p. 7.

26. *Devlin v. Johns-Manville Corp.*, 202 N.J. Super. 556, 495 A.2d 495 (Law Div. 1985).

27. Some courts have permitted epidemiological evidence as a basis for legal proof. For example, in *In re Agent Orange Product Liability Litigation*, *Lilley v. Dow Chemical Co.*, Chief Judge Weinstein noted that the "general scientific technique of inference from animal and other studies is acceptable" (citations omitted). The court recognizes, however, that many epidemiologic studies do not provide results that can be considered sufficiently probative of whether a *particular* plaintiff contracted a disease due to a toxic exposure. This is due to the subjects, methods, and assumptions involved in these studies.

28. *Parker v. Employers Mutual Liability Ins. Co. of Wisconsin*, 440 S.W.2d 43 (Tex. 1969).

29. Ibid., p. 47.

30. *Garner v. Hecla Mining Co.*, 19 Utah 2d 367, 431 P.2d 794 (1967).

31. *Accord Mahoney v. United States*, 220 F. Supp. 823 (E.D. Tenn. 1963), *aff'd*, F.2d 605 (6th Cir. 1964) (plaintiffs failed to establish that workers' exposures to radiation or toxic gases in their employment were sufficient to cause leukemia or Hodgkin's disease); *Lilley v. Dow Chemical Co.* (widow failed to establish that her husband's exposure to Agent Orange caused his lymphosarcoma and coronary artery disease).

32. *Devlin v. Johns-Manville Corp.*

33. Ibid., p. 495.

34. See, for example, *Giordano v. Equitable Life Assurance Society of the U.S.*, 13 Fed. Rules Evid. Serv. 374 (D.N.J. 1983) ( A lay witness may testify as to whether or not the decedent appeared depressed.); See generally, *Richardson on Evidence*, 10th ed. (J.Prince, 1973 and Supp. 1985), sec. 364.

35. Prosser and Keeton, *Torts*, sec. 54; Restatement (Second) of Torts, sec. 436, 436A (1964).

36. Prosser and Keeton, *Torts*, p. 363. See also, *Eagle-Picher Industries, Inc. v. Cox*, 481 So. 2d 517 (Fla. Dist. Ct. App. 1986). A few courts permit recovery for mental distress upon evidence of "impact" even without evidence of present injury. See note 62.

37. See, for example, *Battalla v. State of New York*, 10 N.Y. 2d 237, 176 N.E.2d 729, 219 N.Y.S. 2d 34 (1961).

38. In this passage the authors seem to be referring to both cancerphobia and fear of cancer claims. *Jackson v. Johns-Manville Sales Corp.*, 781 F.2d 394, 414–15 (5th Cir. 1986), quoting Gale and Goyer, "Recovery for Cancerphobia and Increased Risk of Cancer," *Cumberland Law Review* 15 (1985):723, 730–31.

39. *Gideon v. Johns-Manville Sales Corp.*, 761 F.2d 1129 (5th Cir. 1985).

40. *Dartez v. Fibreboard Corp.*, 765 F.2d 465 (5th Cir. 1985), and p. 468.

41. See, for example, *Adams v. Johns-Manville Sales Corp.*, 783 F.2d 589 (5th Cir. 1986); *Ayers v. Jackson Township*, 189 N.J. Super. 561, 461 A.2d 184 (Law Div. 1983), *modified and vacated in part*, 202 N.J. Super. 106, 493 A.2d 1314 (App. Div. 1985).

42. *Jackson v. Johns-Manville Sales Corp.*, p. 412, citing *Entex, Inc. v. Rasberry*, 355 So.2d 1102, 1104 (Miss. 1978); see also *Herber v. Johns-Manville Corp.*

43. *Jackson v. Johns-Manville Sales Corp.* (deciding on Mississippi law); *Herber v. Johns-Manville Corp.*; *Lohrmann v. Pittsburgh Corning Corp.*, 782 F.2d 1156 (4th Cir. Jan. 30 1986) (deciding on Maryland law); *Gideon v. Johns-Manville Sales Corp.* (deciding on Texas law); *Dartez v. Fibreboard Corp.* (deciding on Texas law).

44. *Gideon v. Johns-Manville Sales Corp.* p. 1129. Ibid., p. 1137–38 (citations omitted). *Contra Valori v. Johns-Manville Sales Corp.*, No. 82–2686 (D.N.J. Dec. 11, 1985) (available on LEXIS, Genfed library, Dist. File p. 5), in which the court permitted recovery for increased risk of cancer even though the plaintiff's evidence did not prove that he had a greater than 50 percent chance of contracting cancer in the future. However, this is against the majority of court opinions. The court, allowing the evidence stated:

> Unfortunately, the New Jersey Supreme Court has not yet indicated how the "reasonable probability" standard is to be applied in mass tort situations such as the one here, in which the plaintiff's proof constitutes a showing that he is a member of a class a percentage of whom will definitely incur future harm because of the defendant's tortious conduct, but fails to establish for certain whether he is the person to whom that harm will definitely befall. In the absence of any definitive guidance from the Supreme Court, this court's function is to attempt to predict what its view would be. Defendants suggest that this court should interpret the term "reasonable probability" to mean "more likely than not", i.e., a likelihood greater than fifty percent. According to this reasoning, plaintiffs' claim for damages must fail because their proof establishes only that Mr. Valori is a member of a class forty-three percent of whom will contract cancer.

45. *Arnett v. Dow Chemical Co.*, No. 729586 (Cal. Super. Ct. March 21, 1983) and p. 15 (emphasis in original).

46. *Eagle-Picher Industries, Inc. v. Cox*, 481 So.2d 517 (Fla. Dist. Ct. App. 1986), and p. 524 (citations omitted).

47. *Ayers v. Jackson Township*.

48. Ibid., p. 122 (citation omitted).

49. *Friends For All Children v. Lockheed Aircraft*, 746 F.2d 816, 826 (D.C. Cir. 1984).

50. *Herber v. Johns-Manville Corp.*, *Contra Valori v. Johns-Manville Sales Corp.*; *Devlin v. Johns-Manville Corp.*

51. Prosser and Keeton on Torts, sec. 30, p. 165 (footnote omitted); *Accord Leonhard v. United States*, 633 F.2d 599, 613 (2d Cir. 1980), *cert. denied*, 451 U.S. 908 (1981) ("Under general principles of law, a cause of action accrues when conduct that invades the rights of another has caused injury.") *Amoco Transport Co. v. Bugsier Reederei and Bergungs, A.G.*, 659 F.2d 789, 795 n. 9 (7th Cir. 1981) ("ordinarily a tort of any kind

does not give rise to a cause of action unless and until the plaintiff has suffered harm of a kind legally compensable by damages.''); *Bart v. Telford*, 677 F.2d 622, 625 (7th Cir. 1982) (''A tort to be actionable requires injury.'').

52. *Laswell v. Brown*, 683 F.2d 261 (8th Cir. 1982), *cert. denied*, 459 U.S. 1210 (1983); see also *Westrom v. Kerr-McGee Chemical Corp.*, No. 82-C-2034 (N.D. Ill. Oct. 4, 1983) (the court dismissed a claim for increased risk of cancer resulting from exposure to radiation in a building purchased from the defendant); and p. 269 (emphasis in original) (quoting the district court opinion *Laswell v. Brown*, 524 F.Supp. 847, 850 [W.D. Mo. 1981]).

53. *Ayers v. Jackson Township*.

54. Ibid., pp. 186–87.

55. Ibid., p. 187.

56. Ibid., p. 184.

57. *Arnett v. Dow Chemical Co.*, Slip op. at 15.

58. *Plummer v. Abbott Laboratories*, 568 F. Supp. 920 (D.R.I. 1983); and p. 922 (emphasis in original).

59. *Mink v. University of Chicago*, 460 F. Supp. 713, 719 (N.D. Ill. 1978).

60. *Morrissy v. Eli Lilly & Co.*, 76 Ill. App. 3d 753, 761, 394 N.E.2d 1369, 1376 (1979).

61. *Rheingold v. E. R. Squibbs & Sons, Inc.*, No. 74-CIV-3420 (S.D.N.Y. Oct. 8, 1975).

62. Some jurisdictions do recognize causes of action for *fear* of future cancer even when there is no present injury. Such actions are forms of emotional distress claims. See, for example, *Wetherill v. University of Chicago*, 565 F. Supp. 1553 (N.D. Ill. 1983); *Laxton v. Orkin Exterminating Co., Inc.*, 639 S.W.2d 431 (Tenn. 1982); *Arnett v. Dow Chemical Co.*

63. Claims for property damage have also been brought by plaintiffs who claim exposure to formaldehyde, either from emissions from urea formaldehyde foam insulation or from particle board used in the construction of their residences. It is difficult to assess the resolution in these cases for to date the majority of the claimants are still trying to have their claims certified as class actions. See, for example, *Alley v. Gubser Development Co.*, 785 F.2d 849 (10th Cir. 1986; *Fish v. Georgia-Pacific Corp.*, 779 F.2d 836 (2d Cir. 1985); *Caruso v. Celsius Insulation Resources, Inc.*, 101 F.R.D. 530 (M.D. Pa. 1984).

64. *Corporation of Mercer University v. National Gypsum Co.*, No. 85-126-3-MAC (M.D. Ga. April 11, 1986), *appeal docketed*, No. 8693 (11th Cir. Sept. 22, 1986) (verdict in favor of plaintiffs); *City of Greenville v. W. R. Grace & Co.*, No. 6:85-1693-3 (D.S.C. Jan. 24, 1986), *appeal docketed*, No. 86-2096 (4th Cir. July 11, 1986) (verdict in favor of the plaintiffs); *Spartanburg County School District Seven v. National Gypsum Co.*, No. 83-1744-14 (D.S.C. Aug. 15, 1985), *appeal docketed*, No. 86-2273 (4th Cir. Dec. 2, 1985) (verdict in favor of the plaintiffs); *County of Anderson, Tennessee v. U.S. Gypsum Co.*, No. CIV-3-83-511 (E.D. Tenn. Mar. 12, 1985), *appeal docketed*, No. 85-5474 (6th Cir. May 29, 1985) (verdict in favor of defendants).

A fifth asbestos property damage case went to trial but was settled prior to verdict. *Lexington County School District Five v. U.S. Gypsum Co.*, No. 3:82-2072-0 (D.S.C. Apr. 10, 1984).

65. *County of Anderson, Tennessee v. U.S. Gypsum Co.*

66. Ibid., trial testimony of Dr. Kenny Crump, p. 2475; U.S. Gypsum Co. Trial Exhibit No. 145, "Lifetime Risks Per Million Persons."
67. Ibid., p. 2478, 2479.
68. *Franklin County School Board v. Lake Asbestos of Quebec, Ltd.*, No. 84-AR-5435-NW (N.D. Ala. Feb. 13, 1986), *appeal docketed* No. 86-7170, (11th Cir. Mar. 12, 1986); *Board of Education of the City of Chicago v. A. C. and S., Inc.*, No. 85 CH-00811 (Ill., Cir. Ct. Cook County Feb. 26, 1986), *appeal docketed*, o. 86-817 (Ill. App. Ct. Apr. 1, 1986) (All but one of the fourteen counts in plaintiff's complaint were dismissed. The one count that has not been ruled upon is plaintiff's count regarding punitive damages. The court's order of November 27, 1985, postponed considerations of all issues relating to punitive damages in deference to the mandatory national class which was conditionally certified on the issue of punitive damages for school districts in *Asbestos School Litigation*, 104 F.R.D. 422 (E.D. Pa. 1984). The United States Court of Appeals for the Third Circuit has reversed the mandatory class decision. *In re School Asbestos Litigation, School District of Lancaster v. Lake Asbestos of Quebec, Ltd.*, no. 84-1642 (3rd Cir. May 1, 1986)).
69. *Board of Education of the City of Chicago v. A.C. and S., Inc.*
70. *Gideon v. Johns-Manville Sales Corp.*; see also restatement of Torts, 2d Sec. 910; 22 Am. Jur. 2d Sec. 26.
71. *Gideon v. Johns-Manville Sales Corp.*, p. 1137.
72. Ibid.
73. *Eagle-Picher Industries, Inc. v. Cox*; *Devlin v. Johns-Manville Corp.*; *Wilson v. Johns-Manville Sales Corp.*, 684 F.2d 111 (D.C. Cir. 1982); *Pierce v. Johns-Manville Sales Corp.*, 296 Md.656, 464 A.2d 1020 (1983).
74. *Eagle-Picher Industries, Inc. v. Cox.*
75. Ibid., p. 521.
76. Restatement (Second) of Judgments, Sec. 22(a) (b) and (f) (1980).
77. *Eagle-Picher Industries, Inc. v. Cox.*

# 6
# Trans-Science in Toxic Torts

## WENDY E. WAGNER

Forty years after science established that asbestos was hazardous to humans, the first plaintiff recovered damages against an asbestos manufacturer.[1] This delay resulted from the inability of science to meet the legal, "more probable than not" causation standard that requires that the effect of the hazard first be quantified on humans. A determination of whether exposure to a substance is a quantifiable (or statistically significant) factor in the cause of a disease necessitates an epidemiology study, an inherently retrospective undertaking, in which epidemiologists compare a significant number of subjects exposed to the substance to unexposed populations. Since the first successful epidemiological study was completed, over 20,000 claims have been filed nationwide against manufacturers and installers of asbestos.[2]

This pattern of delay is common to the toxic tort field.[3] In a variety of other cases involving such injuries as adenocarcinoma, pelvic inflammatory disease, toxic shock syndrome, and Guillain-Barre syndrome, the general population was exposed to known health hazards, but recovery was granted only after a statistically significant number of deaths and injuries were incurred, allowing experts to quantify the hazard.[4] In each of these cases, the court demanded epidemiological evidence to satisfy the general causation requirement. Evidence only that the substance was risky (for example, carcinogenic), based on animal and other nonhuman research, was inadequate.

This chapter discusses the inherent problems with a causation standard that requires that the hazardous nature of a substance be quantified in the general population before granting recovery. Scientific quantification requires both that an epidemiological study be conducted, a highly expensive and time-consuming undertaking, and that the study be successful in distinguishing injuries caused by the product from those induced by the general environment.[5] These scientific barriers not only ensure that very few substances will be studied adequately to meet existing legal requirements, but also make it difficult, or even impossible, for a manufacturer to predict liability in the interim period after animal experimentation indicates that a substance is dangerous, but before the hazard is quantified on humans.[6] In order to reincorporate deterrence into toxic tort cases and to provide a basis for determining liability in this scientifically uncertain interim period, the standard for liability must be revised. Such a standard should not only compensate deserving plaintiffs and deter possible future defendants, but should also provide a definitive liability standard that ensures that the "tenuous, fraudulent, or trivial" cases warned of in Chapter 5 do not prevail.

The revised causation standard proposed in this chapter circumvents problems with scientifically indeterminate solutions by combining a qualitative showing of causation with proof that the manufacturer acted negligently in introducing an "unreasonably dangerous" product. Qualitative evidence of a causal link would include proof of substantial exposure to the substance and injury consistent with that substance, rather than the present requirement that plaintiff prove with statistical certainty that causation was "more probable than not." Proof that the substance is "unreasonably dangerous" would involve a finding that a manufacturer acted negligently in marketing a product when it "should have known" that the product posed a serious risk, although still unquantified, to human health. The jury would determine whether the manufacturer's marketing decision that the benefits of the product outweighed the costs to human health was reasonable. If a plaintiff satisfied both of these requirements—proof of a qualitative causal link and the distribution of an "unreasonably dangerous" product—then the burden will shift to defendant to prove that the product was safe, the hazards were not foreseeable, benefits outweighed potential costs at the time of marketing, or that the plaintiff was not exposed to substantial concentrations of the product.

## THE PROBLEM IN TOXIC TORTS: TRANS-SCIENCE

At the heart of the problem presently confronted by the courts in toxic tort suits is the inability to determine causation quantitatively when trans-scientific issues are involved—when questions asked of science, such as the statistically significant effects of a chemical on human health, cannot be answered at the time.[7] For example, early quantification of the risk a chemical poses to human health is impossible because ethical policies preclude tests on a large number of humans.[8] Instead, science is capable only of indicating that a substance is generally dangerous through studies on nonhuman subjects or accidental spills.[9] Thus, for many types of injuries, several decades may pass before harm manifests following low but continuous exposure to the hazardous substance.[10]

In this section, trans-science will be defined and its impact on the judicial system examined. The next section will investigate the legal bases for the present inadequacy of the courts to adjudicate trans-scientific issues. The final section will outline a proposal for reform that might induce the judicial system to adapt to the unique characteristics of trans-science.

## The Source of the Problem

Unlike the traditional scientific process in which hypotheses are constantly refined by experiment and observation, the process involved in trans-science is frozen at an early developmental stage, consisting of a number of competing hypotheses that have never been tested or are tested only superficially and unsystematically.[11] The "trans-scientist" must rely largely on short-term controlled experiments conducted on laboratory animals or isolated accidents to predict the potential risks of exposure in the multivariable outside environment. Accordingly, the realm of trans-science is characterized by an extrapolatory gap that separates the experimentally established effects of a substance on nonhumans or on humans following isolated short-term disasters from the untested predictions of the effects on humans following exposure to low doses over a long period of time.

In the past, trans-science did not appear to pose a problem for courts because most personal injury suits involving the effects of hazardous substances on human health were filed several decades after the substance had been released into the environment. By that time, data had emerged to demonstrate with statistical certainty the magnitude of impact that the substance had on humans, and thus the probability that it caused plaintiff's injury.[12] The most significant trans-scientific issues, therefore, were resolved before the issue ever entered the courtroom. In contrast, many recent toxic tort claims have been filed while the effects of the substance are still clouded by the uncertainties of trans-science—before a statistically significant number of persons have been exposed and a conclusive epidemiological study has been done.[13]

## Impact of Trans-Science on the Judicial System

When determination of the effects of a substance on human health is in this interim, trans-scientific stage, the probability that the substance might be the cause of a disease is unclear. Satisfaction of the traditional "more probable than not" standard becomes impossible, because plaintiffs cannot prove that the substance had at least a 50 percent probability of being responsible for their injuries. Courts generally deny recovery to these plaintiffs and occasionally even refuse to hear their cases. On the other hand, once epidemiological studies are completed and the data indicates that the substance has a statistically significant effect on exposed persons, the problems of trans-science diminish and courts grant recovery in almost every plausible instance.

Figure 6.1 illustrates the dramatic impact the resolution of a trans-scientific

Figure 6.1
The Trans-Science Time Lag

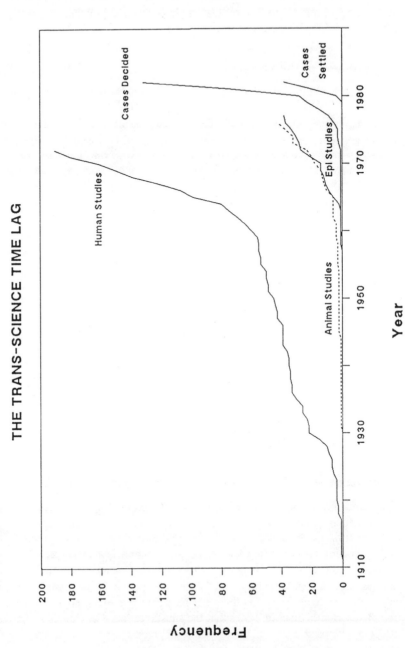

question has on the courts. For decades scientists continued to publish studies indicating the adverse effects of a substance, in this case asbestos, on human health and animals, but the courts and presumably even plaintiffs' lawyers did not act on this information until after the definitive epidemiology study was published in 1965. In fact, an additional five- to ten-year lag time occurred after publication of the study before the legal community apparently discovered or recognized its import. Thus, from this example, it appears that there is an approximate forty-year delay between early scientific recognition of the hazards of a substance and the eventual judicial acknowledgment of those hazards, which in turn results in an immediate, skyrocketing case load for the courts.

The deterrence achieved by those suits that eventually do succeed is clouded considerably by the sequence of events that must occur before the more significant trans-scientific elements are resolved. The uncertainty surrounding the extent of harm a substance will cause humans, the chance that adequate records of exposure will be available, the probability that an epidemiology study will be done, and the probability that it will be successful, together reduce the likelihood that a negligent manufacturer will be found liable. The limits of the scientific process as well as finite resources to study every potential health hazard in a statistically comprehensive way ensure that many hazards will remain undetected and their manufacturers undeterred. Moreover, even when the effects of a substance are detected, compensation may be frustrated by the manufacturer's or insurer's bankruptcy.[14]

The case of *Parker v. Employers Mutual Liability Insurance Company of Wisconsin*, discussed previously in Chapter 5, presents an excellent example of this clouded deterrence, where the uncertainties of scientific detection impede a full assessment of liability.[15] Plaintiff Parker worked as a material handler and production operator, assembling and disassembling nuclear weapons. During his four and a half years of employment, plaintiff was exposed daily to moderate doses of radiation. Although radiation was determined to be highly carcinogenic to animals as early as 1950, there were no conclusive investigations regarding the effects of low dosages on humans. In 1965, plaintiff was afflicted with cancer of the lymph node and sued the allegedly negligent employer. The Supreme Court of Texas affirmed the trial court's judgment for defendant, however. In its holding the court stated that, regardless of what isolated radiation studies might demonstrate, in the absence of a quantitative probability that radiation will cause cancer in humans there was no evidence indicating causation. The judge ignored the fact that he was not ruling on the weight of existing evidence, but rather on the limits of science.[16]

Analyzing the troubling verdicts in the many cases like *Parker*, scholars have focused on the scientific difficulty of determining recovery when several possible causes of an injury exist.[17] To resolve the problem they suggest a proportional liability scheme that calculates damages according to the probability that the injury was caused by the defective product.[18] Such an analysis, however, assumes that the risks can be quantified. It completely overlooks the more onerous, first-order problem confronting courts that arises when science is unable to offer any numerical probability of injury upon exposure to a hazardous substance. Thus, while the problem of sorting out the potential causes of an injury is serious, this chapter examines the more pervasive problem of how courts should handle substantial yet unquantifiable risks.

## TORT SYSTEM FAILURES ON TRANS-SCIENCE ISSUES

There are two major discontinuities between trans-science and the present tort system. First, the statistical requirements for a "more probably than not" standard of causation demand a certainty in quantifying causation that trans-science is incapable of producing. Second, the rules of evidence and procedure are incompatible with the capabilities of trans-science.

### Trans-Science Incompatibility with Causation Standards

The standard for proof of causation in toxic tort cases requires that the occurrence be large enough to make it "more probable than not" that an individual plaintiff's injury resulted from a hazard produced by defendant.[19] This traditionally involves two types of scientific proof: (1) epidemiology studies indicating that the risk that a hazard will cause a specific injury is at least twice the normal background level of risk for that injury, and (2) medical proof that links defendant's hazard with each of plaintiff's injuries.[20] Satisfaction of these requirements is generally sufficient to shift the burden to defendant, even if plaintiff's claim did not preclude all other causes.[21]

In cases involving trans-science, resolution of the first element is often decisive. While experimentation on animals and other organisms may indicate a plausible relationship between exposure to a substance and resulting injury, the courts' further quantitative requirement that the occurrence of the disease in those exposed to a substance be twice the background incidence creates an insurmountable obstacle in trans-scientific cases. This mandate essentially requires direct experimentation on humans, with a sample size large enough to yield statistically significant results. Consequently, when courts impose "more probable than not" or "but for" causation standards on trans-scientific issues, failure is inevitable, because strong probabilistic evidence for causality is being demanded from scientists who are unable to conduct the necessary experiments.

The case of *Johnson v. United States* illustrates this anomalous result.[22] In *Johnson*, plaintiffs alleged that their cancers were a result of daily occupational exposure to minute amounts of ionizing radiation. In denying plaintiffs recovery, the court refused to interpret the Kansas causation standard requiring proof to meet a "reasonable degree of medical certainty" as only meaning that the bulk of scientifically available data support plaintiffs. Instead, the court held that the specification necessitated "medical certainty" that plaintiffs' exposure to radiation was the cause of their resulting cancer—an imperative well beyond the capability of contemporary science. The court's words are revealing: "We can see that in matters of determining the cancer risks from low occupational doses of radiation, scientists do not deal with what exists in fact and can be measured or experimentally proven. . . . Cause in tort law needs to be founded on more than a theory or hypothesis."[23]

Several judges have recognized that the traditional causation requirements are incompatible with trans-science and have responded by applying a weaker, qualitative standard.[24] Because most courts have adhered to some form of "more probable than not," however, an inconsistency between courts has ensued, further eroding any predictable basis for liability.

## Trans-Science Incompatibility with Traditional Judicial Constraints

A second problem emerges when courts are forced to evaluate and interpret trans-scientific issues under existing evidentiary and procedural rules.

*Evidentiary Standards of Admissibility.* In the courts the adversarial process seeks to determine the truth from various sets of facts collected by adverse parties. The Federal Rules of Evidence were developed to distinguish necessary and relevant facts from those that are unnecessary. Unfamiliar with the terminology and principles of science, judges are nevertheless forced to make subtle distinctions between fact and nonfact, and between theories that are generally accepted within the scientific community and those that are controversial. In cases involving trans-science, precisely these questions cannot be answered by the scientific community itself. Without definitive data or experiments, there is no generally accepted or reasonable scientific basis for substantiating the finding of a causal connection, and the use of indirect extrapolations from animal studies may not be considered relevant to the effects on humans.[25] Thus, disputes over the admissibility of evidence center on the scientific validity of vying hypotheses rather than on the legal relation of relevant facts to facts that are irrelevant or have only minimal probative value.

Judicial opinions illustrate the courts' wildly varying views on the "factual" value of trans-science hypotheses. In *Lima v. United States* the court held expert testimony to be inadmissible under Rules 702 and 703 of the Federal Rules of Evidence because the "smoldering Guillain-Barre Syndrome (GBS)" theory used to explain plaintiff's injury was not of the type "reasonably relied on by experts in the field."[26] Specifically, the court noted that while the theory demonstrated the causal connection between exposure to the swine flu immunization shot and resulting injuries consistent with that exposure, the factual basis to support the theory was simply too incomplete—a consequence of the small data base, not of the scientist's lack of credibility or methods of experimentation. Due to this inadequate factual basis, the court refused to grant recovery. In several swine flu cases, however, courts have accepted the "smoldering GBS" theory and have granted relief to plaintiffs with claims similar to Lima's.[27]

Likewise, in *In Re Agent Orange*, Judge Weinstein employed the Federal Rules of Evidence to remove much of the science that he considered either "irrelevant" or "conclusory," which included all studies done on animals or arising from isolated industrial accidents.[28] While one of plaintiff's experts withstood a motion challenging his credibility, the court held his theories to be inadmissible under Rule 403. Although the court did note that the evidence might mislead "at least some members of the jury," the court based its holding more solidly on the desire to put "this prolonged litigation" to an end.[29] "In complex and protracted litigation, waste of the trier's time is a particularly telling factor."[30] In affirming Judge Weinstein's ruling, the Second Circuit acknowledged the difficulty of proving causation, but nevertheless reinforced Judge Weinstein's blanket dismissal of animal and industrial studies:

Studies based on industrial accidents and experiments with animals suggest that exposure to dioxin may cause various of those ailments. However, these studies involve different dosages and different species than are involved in this litigation. Studies of Vietnam veterans themselves fail to demonstrate ailments occurring among them at a statistically abnormal rate. The weight of present scientific evidence thus does not establish that personnel serving in Vietnam were injured by Agent Orange.... [although] [s]uch studies are, of course, not conclusive.[31]

The Federal Rules of Procedure raise even more onerous difficulties in the evaluation of trans-scientific issues. It is the general rule that summary judgment, judgment notwithstanding the verdict, and judgment on the pleadings are granted only when there is no material issue of fact in dispute. When confronted with disputes over interpretation of uncertainty, courts are generally reluctant to grant summary judgment, limiting its use to situations in which the facts are based on more *pro forma* denials, sham, or patently false assertions in the pleadings. These patterns are not followed uniformly in the toxic tort area, however.[32] Judge Weinstein's ruling that animal studies and over one hundred privately conducted epidemiological studies were inadmissible heavily influenced the court's determination that there were no facts in dispute in granting defendant's motion for summary judgment, because most of plaintiffs' remaining evidence relied, at least in part, on extrapolations from these excluded studies.[33]

In sum, under our present system judges are called on to make scientifically delicate determinations of what is resolvable, certain, or factual in order to quantify causation under legal rules designed for more determinable situations. If lines must be drawn, they will be drawn by judges who often ignore the inherently limited capabilities of scientific research, leading inevitably to inconsistent and haphazard judgments. Such inconsistency, in turn, precludes adequate financial planning for liability by manufacturers when initially marketing the product, and ultimately impairs the ability of the courts to deter wrongful conduct. Instead of providing a predictable basis for imposing liability, manufacturers will conclude that liability is based on each judge's personal perception of "fact" and "relevancy."

## PROPOSAL FOR REFORM

In order to provide proper incentives for deterring wrongful behavior and to ensure equitable compensation to injured victims in toxic tort cases, a standard for liability must be devised that can be applied consistently and clearly. The standard must adjust to the inability of trans-science to quantify the effects of a substance. It must also resolve or circumvent the evidentiary and procedural problems resulting from the inherently hypothetical rather than factual nature of trans-science. The reform proposed here would shift the burden of proof to the defendant following (1) proof of qualitative causation, and (2) proof that the product was "unreasonably dangerous." Together, these requirements should ensure that plaintiffs will have their cases heard, while the manufacturers will have a consistent basis for predicting future liability.

Reincorporating notions of negligence while simultaneously weakening the necessary proof of causation may cause earlier imposition of liability—before the risk is tested on humans—to the disadvantage of manufacturers. But many manufacturers may prefer a more predictable basis for risk. Perhaps even at the cost of imposing liability earlier. Additionally, since manufacturers often are both in more informed positions and are imposing their products on involuntary, nonpurchasing plaintiffs, there is potentially less cost in attaching greater burdens on defendant manufacturers. Finally, fairness demands that a risk of error falling on plaintiffs should be partially shifted to defendants because, at present, a substantial number of deaths are necessary before liability is imposed.

### Qualitative Causation

Under the present "more probable than not" causation requirements, plaintiffs in toxic tort cases are unable to sustain their burden of proof because the resulting injuries have not been statistically detected by epidemiologists. The procedural burden of proof, however, should be refocused on the underlying realities of the typical toxic tort case. Plaintiffs injured by hazardous products must have the power to be heard, despite the fact that the statistical impact of the hazard has not yet been determined. The revised causation standard should be based, then, on more qualitative elements of causation, such as proof that the plaintiff was exposed to substantial concentrations of the substance and was afflicted with an injury consistent with the known hazards of that substance.[34] In fact, this qualitative causation standard, standing alone, was used to shift the burden of proof to defendants by the district court in *Allen* and has been introduced in Congress as a statutory reform proposal.[35]

This qualitative standard, unlike its present quantitative counterpart, utilizes all available information regarding the hazardous substance, as well as extrapolatory techniques, to determine potential effects on humans. The difficulty of resolving trans-scientific problems is also eased by a requirement that plaintiffs need only prove that a substance is capable of causing an injury consistent with their injury. This burden may be satisfied by a thorough search of studies on the effects of the substance on animals or on humans in occupational settings and following disasters.

Focusing litigation on the qualitative, resolvable aspects of causation also avoids the evidentiary problems presented by trans-science. Under the present "more probable than not" standard, admissibility is dependent upon (1) whether each court wishes to view unverified hypotheses of a substance's effect on humans as accepted within the scientific community, and (2) whether extrapolations from the effects of a substance on animals are relevant to humans. Under the proposed qualitative standard, however, evidentiary questions will involve more commonplace disputes over the facts related to exposure and to diagnosis of injuries in plaintiff that are consistent with the known hazards of the substance. Similarly, procedural rulings will be based on the weight of all evidence, unlike the present rulings that consider only the scant evidence that remains after all nonhuman and inconclusive or irrelevant human studies have been ruled inadmissible.

### Unreasonably Dangerous

Under the existing strict liability standard, relaxing plaintiffs' burden of proof on causation, however necessary from the standpoint of fairness, would subject manufacturers to inordinate liability. Manufacturers would be overdeterred and held accountable for injurious effects that were undetectable when the products were marketed. Hence, before the burden of proof shifts to defendant, plaintiffs ought to be required to prove that the product was "unreasonably dangerous" in light of the manufacturer's knowledge at the time of marketing.[36] The standard suggested here employs "unreasonably dangerous" as a gauge for unreasonable conduct.[37] In short, the two-tiered requirement for liability based on qualitative causation and unreasonable conduct ensures that incentives to create necessary products for human use are properly balanced with the need to deter the manufacture of unsafe products. The "unreasonably dangerous" standard rests on foreseeability of harm and thus incorporates traditional negligence concepts.[38] First, the jury must determine whether a hazardous substance was marketed after the manufacturer "should have known" of its hazardous nature.[39] If the jury confirms such knowledge, it will then employ more traditional strict liability concepts to determine liability, weighing the costs of the uncertain hazardous risk that the product presented at the time of marketing against its apparent benefits to society. The jury will proscribe a product as "unreasonably dangerous" if it makes both findings: (1) the manufacturer "should have known" of the hazardous nature of a product; and (2) the costs of the hazard outweighed its benefits to society at the time of marketing.[40]

Although the cost/benefit approach to "unreasonably dangerous" requires case by case inquiry, in most cases juries are likely to weigh the quantity and quality of animal studies and disaster or occupational reports existing at the time of marketing against the economic and social utility of the product at that time. These determinations do not involve complex numerical weighing, but require only that the jury consider qualitatively which sorts of risks society wishes to assimilate and which it chooses to deter.

Because it relies on deterring knowable but unreasonable risk, this "unreasonably dangerous" standard also comports with traditional cost-spreading notions. Manufacturers will be held liable for damages, and thus have financial incentives to internalize the costs of the unreasonable and preventable risks that their products create. If, on the other hand, the jury determines that the benefits of the product are considerable, the product may still be marketed with adequate warnings to alert the user. When the potentially dangerous nature of the product is within reasonable contemplation and knowledge of the user, however, the product may not need a warning.[41]

This "unreasonably dangerous" standard will also avoid the evidentiary problems posed by trans-science. The current focus by courts on the viability of competing trans-scientific hypotheses, which attempt to quantify the untested effects of a substance on humans, will shift to negligence questions of unreasonable conduct. Moreover, such conduct will be adjudicated in light of the published literature and the manufacturer's satisfactory completion of necessary tests—issues that are determinable. As a result, deterrence will be enhanced in a predictable manner by shifting the burden of proof to manufacturers who negligently market hazardous products, despite the trans-scientific inability to quantify such hazards with certainty.

In sum, basing liability on a qualitative showing of causality and on proof that the manufacturer acted negligently in marketing an "unreasonably dangerous" product will provide a firm basis for deterrence while conforming to traditional tort requirements. It should also be noted that although the proposed liability standard is intended to address trans-scientific problems, its use need not be so limited.[42] Revising the criteria upon which a shift in the burden of proof is based ensures that negligent manufacturers must prove that a trans-scientific substance is safe. Thus when the issue is no longer trans-scientific and the relevant epidemiological studies have been done, or when the manufacturer is able to prove that the product is not hazardous, that the product did not cause plaintiff's injury, or that the plaintiff was contributorily negligent, then the manufacturer may rebut the presumption and shift the burden of proof back to the plaintiff. Although this process may involve an additional burden shift in cases where the most significant trans-scientific elements have been resolved, it does not affect the ultimate basis for imposing liability.

## CONCLUSION

Trans-science in toxic torts presents an obstacle in traditional tort litigation that not only produces confusing and inconsistent judgments, but also makes it difficult or impossible to preserve the goal of deterrence. In order to accommodate trans-science, the judicial framework must change. A proposal for reform is suggested in which trans-scientific obstacles can be circumvented by referring to more predictable notions of qualitative causation and unreasonable conduct. By adopting such a proposal, the courts may be able to reincorporate the principle of deterrence into the adjudication of toxic torts.

## NOTES

The opinions and ideas expressed in this chapter do not necessarily reflect those of the Justice Department. An earlier version of this chapter was published in *Yale Law Review* 96 (1986): 428.

1. Cases of asbestosis in asbestos textile workers had been reported as early as 1924; see Cooke, "Fibrosis of the Lungs Due to the Inhalation of Asbestos Dust," *British Medical Journal* 2 (1924): 147.
2. Selikoff and his colleagues adduced a definitive quantitative assessment of the risk of contracting asbestosis following exposure to asbestos in their seminal 1965 study. I. J. Selikoff, J. Churg, and E. C. Hammond, "The Occurrence of Asbestosis Among Industrial Insulation Workers," *Annals of the New York Academy of Science* 132 (1965): 139. The authors examined 1,522 members of an insulation workers union in the New York/New Jersey area and discovered that almost half of those examined exhibited signs of pulmonary asbestosis. In the subgroup of workers employed over forty years, abnormalities were detected in over 90 percent. By mid–1982, 11,000 health-related cases brought by 15,500 plaintiffs were pending against Johns-Manville Corporation alone. Thereafter, suits were filed at a rate of about 425 cases per month. See W. Lundquist, "Innovations in Mass Tort Litigation" (paper delivered at the Meeting of the American Association of Law and Science, Section on Torts, January 23, 1984). For the effect this sudden burst of litigation had on the judicial system, see Figure 6.1.
3. In this chapter "toxic tort" will be defined as broadly as possible to include cases in which (1) a considerable lag time has occurred between exposure and the evidence of disease, (2) multiple causes explain the onset of a disease, and/or (3) both exposure and the number of possible causes are uncertain. For a more thorough discussion of the types of toxic torts, see Strock's presentation in Chapter 9.
4. See, for example, *Sindell v. Abbott Laboratories*, 26 Cal. 3d 588, 607 P.2d 924 (1980); *Abel v. Eli Lilly & Co.*, 418 Mich. 311, 343 N.W.2d 164 (1984); *Coursen v. A. H. Robins Co.*, 764 F.2d 1329 (9th Cir. 1985); *Kehm v. Procter & Gamble Mfr.*, 724 F.2d 613 (8th Cir. 1983); and *In re Swine Flu Immunization Products Liability Litigation v. United States*, 533 F.Supp. 567 (D. Colo. 1980), *aff'd.* 708 F.2d 502 (10th Cir. 1983).
5. See, for example, H. Northrup, R. Rowan, and C. Perry, *The Impact of OSHA: A Study of the Effects of the Occupational Safety and Health Act on Three Key Industries—Aerospace, Chemicals and Textiles* (Ann Arbor, Mich: Books on Demand, 1978) p. 232. For example, epidemiological data is available for only 14 percent of the more than 400 chemicals currently reviewed by the International Agency for Research on Cancer.

See R. Althouse et al., "An Evaluation of Chemicals and Industrial Processes Associated with Cancer in Humans Based on Human and Animal Data: IARC Monographs Volumes 1 to 20," *Cancer Research* 40 (1980): 1–2.

6. Although Congress has supplemented torts liability with statutory and regulatory controls on hazardous substances, these legislative remedies have proven largely ineffectual. First, statutory control of toxic substances is dispersed over a dozen statutes and regulated by five agencies. This piecemeal regulation produces inconsistent and inefficient results. Second, the small budgets of these agencies preclude effective regulation of a vast number of chemicals. Third, the regulatory agencies tend to operate from a set of static rules that do not promote technological innovation or adapt to improvements in research techniques. See J. Trauberman, "Statutory Reform of Toxic Torts: Relieving Legal, Scientific, and Economic Burdens in the Chemical Victim." *Harvard Environmental Law Review* 7: (1983) 177, 203–205. Finally, many toxic substances simply fall between the statutory cracks and are not subject to direct regulation. See B. Furrow, "Governing Science: Public Risks and Private Remedies," *University of Pennsylvania Law Review* 131 (1983): 1403.

7. Alvin Weinberg, a nuclear physicist at Oak Ridge National Laboratory, first developed the notion of "trans-science":

> The point missed . . . is that the seemingly simple question "What is the effect on human health of very low levels of physical insult?" can be stated in scientific terms; it can, so to speak, be asked of science, yet it cannot be answered by science. I have . . . proposed the name *trans-science* for such questions that seemingly are part of science yet in fact transcend science. . . . [Even] any null experiment—that is, an experiment that shows no biological effect at low levels of insult—does not *prove* the insult is harmless, since a larger experiment might show effects. . . . I must stress that where low-level effects are concerned, there will always be a trans-scientific residue. To decide on standards when science can say neither yea nor nay requires some procedure other than the one usually used by scientist in resolving bona fide scientific questions.

Letters to the Editor, *Science* 174 (1971): 546–47 (letter from Alvin Weinberg); see also Weinberg, "Science and Trans-Science," *Minerva* 10 (1972): 209.

In this crystal-ball question, the tools that would be necessary for an accurate prediction would include the ability to test hazardous substances on a large number of human subjects in carefully controlled circumstances over a long period of time. In the case of the effects of Agent Orange, for example, the definitive study would examine the effects of the herbicide on large populations at various low exposures over a period of twenty to fifty years. In addition, an epidemiologist would have to isolate control populations identical to the exposed populations to eliminate the effects of all other variables. Resulting statistics might indicate correlations between exposure to Agent Orange and resultant injuries.

8. See Charles Fried, *Medical Experimentation: Personal Integrity and Social Policy* (New York: Elsevier, 1974); National Academy of Sciences, *Experiments and Research with Humans: Values in Conflict* (1975); Robert J. Levine, *Ethics and Regulation of Clinical Research* (Baltimore: Urban & Schwarzenberg, 1981); and World Health Organization, *Principles and Methods for Evaluating the Toxicity of Chemicals*, pt. 1 (1978), pp. 41–43.

9. See William D. Rowe, *Evaluation Methods for Environmental Standards* (Boca Raton, Fla: CRC Press, 1983), pp. 26–29. Most experts agree that tests on animals, short-term tests on microorganisms, and chemical structure analyses generally provide the best, if not the only available information about the tendency of a substance to cause chronic health impairments in humans. See Work Group on Risk Assessment, Interagency Regulatory Liaison Group (IRLG), "Scientific Bases for Identification of Potential Car-

cinogens and Estimation of Risks," *Federal Register* 44 (1979): 39, 862–71; Edward J. Gralla, "Protocol Preparation: Design and Objectives in Scientific Considerations," in *Scientific Considerations in Monitoring and Evaluation Toxicological Research*, ed. E. Gralla (Washington: Hemisphere, 1981), pp. 1–2; G. Vettorazzi, *Handbook of International Food Regulatory Toxicology Profiles*, 1 (1981); and World Health Organization, *Principles and Methods for Evaluating the Toxicity of Chemicals*, p. 74.

10. It often takes five to forty years for a disease to appear after continuous, low dose exposure to a hazardous substance. During this time, the victim may be exposed to other variables that may cause a similar disease or symptoms. See J. Kelsey, W. Thompson, and A. Evans, *Methods in Observational Epidemiology* (New York: Oxford University Press, 1986) pp. 14–16; Hall and Silbergeld, "Reappraising Epidemiology: A Response to Mr. Dore," *Harvard Environmental Law Review*, 7 (1983): 441; see also J. Rodricks and R. Tardiff, "Conceptual Basis for Risk Assessment," in *Assessment and Management of Chemical Risks*, ed. J. Rodricks and R. Tardiff (Washington: American Chemical Society, 1984), pp. 3–4. Additionally, records of exposure may be lost in intervening years, which further complicates attempts to correlate exposure with the onslaught of a disease.

11. The degree to which an issue is trans-scientific does vary, however. First, not all substances considered in toxic tort cases are equally trans-scientific, and some are not trans-scientific in any way. The molecular structure of a chemical or the existing literature documenting a chemical's impact on health determines, to some extent, the level of certainty regarding that substance's effect on humans. Second, whether an issue is more or less trans-scientific depends on the nature of the injury inflicted. Substances that inflict diseases not specific to that particular chemical but instead have a high probability of natural occurrence are more trans-scientific than substances that inflict specific diseases. Asbestos and DES injuries are unusual and specific to their particular substances, thus proof of mesothelioma or adenocarcinoma, coupled with evidence of exposure to the substance, indicates an almost undeniable link. Most other substances, however, such as Agent Orange, radiation, indoor air pollutants, and hazardous wastes inflict injuries that are common in the everyday world and therefore difficult to trace to any specific cause. Finally, the substantial lag time before any injury becomes apparent—typically several decades—adds still another trans-scientific component. In addition to the legal problems that statutes of limitations present, often the evidence, data, and other records regarding the duration or extent of exposure necessary to establish a causal relation are lost.

12. See, for example, *Ellis v. International Playtex, Inc.*, 745 F2d 292 (4th Cir. 1984) where a husband commenced action against a tampon manufacturer for the death of his wife resulting from toxic shock syndrome; *Borel v. Fibreboard Paper Products Corporation*, 493 F.2d 1076, 1083–84 (5th Cir. 1973), an action against an asbestos manufacturer for asbestosis; *Lima v. United States*, 508 F. Supp. 897 (D. Colo. 1981), *aff'd* 708 F.2d 502 (10th Cir. 1983), where plaintiff bought suit to recover for injuries allegedly resulting from swine flu inoculation; and B. Black and D. Lillienfield, "Epidemiologic Proof in Toxic Tort Litigation," *Fordham Law Review*, 52 (1984): 732, 773–75, which traces swine flu case law with relevant epidemiology studies.

13. See, for example, *In re Agent Orange Product Liability Litigation*, 611 F. Supp. 1223 (E.D.N.Y. 1985), *aff'd.* 818 F.2d 187 (2d Cir. 1987), where Vietnam veterans not participating in the class action brought suit against a manufacturer of Agent Orange for injuries resulting from spraying; *In re Agent Orange Product Liability Litigation*, 597 F. Supp. 740 (E.D.N.Y. 1984), *aff'd.* 818 F.2d 145 (2d Cir. 1987), the class-action suit

that veterans brought against a manufacturer of the herbicide Agent Orange for injuries allegedly resulting from spraying of Agent Orange in Vietnam; *Johnson v. United States*, 597 F. Supp. 374 (D. Kan. 1984), where four employees of an aircraft instrument plant brought suit against the government for cancers allegedly resulting from exposure to radiation in luminous dials; and *Allen v. United States*, 588 F. Supp. 247 (D. Utah 1984), *rev'd on other grounds*, 816 F.2d 1417 (10th Cir. 1987), where twenty-four plaintiffs brought suit against the U.S. government to recover for cancers allegedly resulting from the testing of atomic weapons.

In affirming the district court in *In re Agent Orange*, 818 F.2d 187, 193, the Second Circuit provided a comprehensive summary of trans-science, although the court's acknowledgment of the problem appears to contradict its subsequent judgment granting defendant manufacturer's motion for summary judgment: "While the decisions to use Agent Orange were being made, the most relevant question was not, 'What will dioxin do to animals?' or even, 'What will dioxin do to humans exposed to it in industrial accidents?' The most relevant question was, 'What will Agent Orange do to friendly personnel exposed to it?' The epidemiological studies ask the latter question in hindsight and answer, 'Nothing harmful so far as can be told.'"

14. See R. Epstein, "The Legal and Insurance Dynamics of Mass Tort Litigation," *Journal of Legal Studies* 13 (1984): 475, 495–505, which discusses problems of insurance coverage in mass tort cases. Note as well, "The Manville Bankruptcy: Treating Mass Tort Claims in Chapter 11 Proceedings," *Harvard Law Review* 96 (1983): 1121.

15. 440 S.W. 2d 43 (Tex. 1969).

16. Ibid., p. 49. The court did note: "This requirement does in some instances place extraordinary burdens of proof on claimants. But once the theory of causation leaves the realm of lay knowledge for esoteric scientific theories, the scientific theory must be more than a possibility to the scientist who created it."

17. For example, Rosenberg observed: "Mass exposure cases present two distinct varieties of specific-causation questions. First, it is often unclear which one of several manufacturers of a given toxic agent produced the particular unit of the substance that harmed the plaintiff. Second, and far more common, is the problem of determining the origin of the victim's disease." D. Rosenberg, "The Casual Connection in Mass Exposure Cases: A 'Public Law' Vision of the Tort System," *Harvard Law Review* 97 (1984): 849, 856; see also Black and Lilienfeld, "Epidemiologic Proof in Toxic Tort Litigations," p. 750.

18. See, Rosenberg, "The Causal Connection," p. 859, where the relationship between proportional liability and probability of causation is discussed.

19. General standards for "more probable than not" include a showing of: "(1) exposure significant enough to trigger disease; (2) a demonstrated, biologically plausible relationship between the chemical and disease; (3) the diagnosis of such disease in the plaintiff; and (4) expert opinion that the plaintiff's disease was was [sic] consistent with exposure to the chemical." See Hall and Silbergeld, "Reappraising Epidemiology: A Response to Mr. Dore," p. 445.

20. An even stronger version of this causation standard was employed in *Johnston v. United States*, 597 F. Supp. 374 (D. Kan. 1984), where the court required proof of causation to a medical certainty.

21. See, for example, *American Life Ins. v. Moore*, 216 Ark. 44, 223 S.W. 2d 1019 (1949), where a jury award of death benefits under employee group accident policy was affirmed even though medical experts admitted fatal pulmonary embolism might not have

been caused by previous work-related injury; *Smith v. Humboldt Dye Works, Inc.*, 34 A.D. 2d 1041, 312 N.Y.S.2d 284 (1970), which affirmed Workmen's Compensation Board's award of benefits to an employee engaged in dying wool yarns, even though medical experts could not definitively link the employee's bladder cancer to exposure to aniline dyes.

22. 597 F.Supp. 374 (D. Kan. 1984).

23. Ibid., p. 425. Although a slightly weaker burden of proof was required to establish causation in *In re Agent Orange Product Liability Litigation*, the result was the same. Plaintiffs were asked to produce epidemiological studies, which could not be obtained, due primarily to inadequate records of exposure, in order to quantify the probability that plaintiffs' injuries were caused by the herbicide: "No acceptable study *to date of Vietnam veterans and their families* concludes that there is a causal connection" [emphasis added].

The case of *Stites v. Sundstrand Heat Transfer, Inc.*, 660 F. Supp. 1516 (W.D.MI. 1987), presents an even more egregious example of the illogical results arising from application of an inappropriate causation standard. In *Stites*, residents brought suit against the owners of a manufacturing plant alleging that exposure to the toxic chemicals from the plant caused both an increased risk of future cancer and severe emotional distress resulting from the increased risk. The district court granted defendant's motion for partial summary judgment based largely on plaintiffs' inability to meet the Michigan standard that requires proof that the future cancer is "reasonably certain" to occur: "Given the standard for recovery, the Court does not believe that plaintiffs have met their burden of designating specific facts showing that there is a genuine issue for trial.... First, the Court notes that none of the plaintiffs' experts were able to quantify the enhanced cancer risk plaintiffs face because of their exposure to TCE. The Court appreciates that difficulty of quantifying plaintiffs' enhanced risk of cancer, particularly given that plaintiffs were exposed to several other chemicals that also may be carcinogenic. Yet plaintiffs were unable to establish that they have anything near a reasonable certainty of getting cancer in the future. Given this lack of significant probative evidence on this portion of plaintiffs' claims, the Court must find that plaintiffs have failed to establish that there exist material factual issues to be tried." [Pp. 1524–25.]

24. See, for example, *Allen v. United States*, 588 F. Supp. 247 (D. Utah). The court stated: "Judges and lawyers must approach with great care, the idea that court decisions can be justified solely on the findings of science, lest the quest for justice be lost along the way [quoting H. T. Markey, "Needed: A Judicial Welcome for Technology," 79 *Federal Rules Decisions* (1979): 209, 211].... In the pragmatic world of 'fact' the court passes judgment on the probable. Dispute resolution demands rational decision, not perfect knowledge" (p. 260 ). Unfortunately, however, in reviewing the case on appeal, the Tenth Circuit reversed the district court's judgment on other grounds, unrelated to causation. See *Alan v. United States*, 816 F.2d 1417.

The Eleventh Circuit has also recognized the serious problems caused by trans-science in toxic tort cases, but has taken a somewhat different approach. In *Wells v. Ortho Pharmaceutical Corp.*, 788 F.2d 741, 744 (11th Cir. 1986), plaintiffs alleged that a spermicide caused birth defects in the infant daughter. Noting first that " 'products liability law [should]... not preclude recovery until a 'statistically significant' number of people have been injured,' " the court concluded, "the district court properly noted that 'its ultimate focus was *the birth defects suffered by Katie Wells*. Plaintiffs' burden of proving that Katie Wells's defects were caused by the product did not necessarily require them to produce scientific studies showing a statistically significant association between spermicides and congenital malformations in a large population.' "

25. In *Johnson v. United States*, 597 F. Supp. 374, 409-15 (D. Kan. 1984), for example, the court rejected the testimony of two expert witnesses because they had no reliable or significant evidence on which to premise their opinions; neither served on national committees chosen to address the question under dispute; neither had accepted the consensus reports of the committees as reliable authorities; and both used unreliable statistical methods not commonly used in that particular field of science. The court's conclusion was abrupt: "Thus, when the doses are so low that the existence of any effect at all is only hypothetical theory, such calculations should not, nor will they, be accepted here as valid evidence on causation" (p. 426). "A theory of hypothesis or assumption which yields a number like 97.6% or 8% is not yielding a real number" (p. 425).

In contrast, the court in *Kehm v. Procter & Gamble Mfr.*, 724 F.2d 613, 618 (8th Cir. 1983), admitted government epidemiological studies on toxic shock syndrome, in spite of Procter and Gamble's challenges that they were not "factual findings," not done by persons with first-hand knowledge in the field, and untrustworthy. The court ruled against exclusion on several grounds: the procedures were widely accepted in the field of epidemiology, the investigations were timely and objective, and the individuals preparing them were especially skilled. "There is no reason not to admit the findings simply because they tend towards the conclusory rather than the factual end, unless ... the sources of information or other circumstances indicate lack of trustworthiness" (p. 618, citing *United States v. American Tel. & Tel. Co.*, 498 F. Supp. 353, 360 [D.D.C. 1980]).

26. 508 F. Supp. 897 (D. Colo. 1981), *aff'd.* 708 F. 2d 502 (10th Cir. 1983). In addition, the court in *Beighler v. Kleppe*, 633 F. 2d 531 (9th Cir. 1980), gave insight into the traditional application of these Federal Rules of Evidence: "Rules 702 and 704 allow properly qualified experts to testify in the form of an opinion about issues as to which their expertise may assist the trier of fact, even if the opinion embraces an ultimate issue of fact. Rule 703 permits the expert to base opinions or inferences on facts or data not admissible in evidence if they are of a type reasonably relied upon by experts in the field. Rule 705 permits an expert to give opinion testimony without prior disclosure of the underlying facts or data" (p. 533).

27. See, for example, *Barnes v. United States*, 525 F. Supp. 1065 (M.D. Ala. 1981); see also *Spencer v. United States*, 569 F. Supp. 325 n.2 (W.D. Miss. 1983).

28. 611 F. Supp. 1223 (E.D.N.Y. 1985); and, "A number of sound [governmentally conducted] epidemiological studies have been conducted on the health effects of exposure to Agent Orange. These are the only useful studies having any bearing on causation. All the other data supplied by the parties rests on surmise and inapposite extrapolations from animal studies and industrial accidents." *In re Agent Orange*, p. 1221. Weinstein did not support his decision to exclude all epidemiology studies not conducted by the government except by asserting that some of the studies relied on inapposite data or were flawed (p. 1241). The exclusion of all animal studies were based on the fact that the concentrations used in the animal studies were higher than those in the environment and because the animal studies "involve different biological species." Judge Weinstein ignored the fact that extrapolation from animal studies is a widely utilized scientific method and comprised the primary evidence available on the health effects of Agent Orange.

29. Ibid., p. 1256.

30. Ibid. Since deciding the Agent Orange case, Judge Weinstein, along with other scholars, has advocated that trial judges play an even more active role in eliminating scientific evidence from the jury's consideration. See J. Weinstein, "Litigation and Statistics: Obtaining Assistance Without Abuse," *Toxics Law Report* (BNA) 1 (1986):

182; Zata and Sherwood, "Defending Speculative Injury Claims," *Toxics Law Report* (BNA) 2 (1987): 76; Institute for Health Policy Analysis, Georgetown University Medical Center, *Causation and Financial Compensation* (1986); and Black "Evolving Legal Standards for the Admissibility of Scientific Evidence," *Science* 239 (1988): 1508.

31. *In re Agent Orange Product Liability Litigation*, 818 F.2d 145, 172 (2d Cir. 1987). See also *Lynch v. Merrell-National Laboratories*, 830 F.2d 1190, 1194 (1st Cir. 1987), where the court stated "*in vivo* animal studies, and the study of 'analogous' chemicals . . . singly or in combination, do not have the capability of proving causation in human beings in the absence of any confirmatory epidemiological data."

32. See, *Stiles v. Union Carbide Corporation*, 520 F. Supp. 865 (S.D. Tex. 1981), wherein summary judgment was granted to defendants in a wrongful death resulting from exposure to toxic chemicals because evidence of defendant's concealment of hazardous nature of chemicals was insufficient to toll statute of limitations; *Synalloy Corporation v. Newton*, 254 Ga. 174, 326 S.E.2d 470 (1985), where employees brought suit against their employer for negligent exposure to carcinogenic betanapthylamine—the court granting summary judgment to defendant employer because disabilities had not yet manifested and Georgia statute of limitations barred employee claims one year following exposure.

33. 611 F. Supp., pp. 1259–60. Judge Weinstein's unsupported and seemingly harsh ruling contrasts with traditional features of conclusory allegations: unsupported evidence (no connection between fact and allegations); incomplete evidence; contradiction of known facts that both parties agree on; and internally inconsistent claims. See, for example, *Cunningham v. Rendezvous, Inc.*, 699 F.2d 676, 678 (4th Cir. 1983).

34. In fact, the *Restatement (Second) of Torts* mentions causation, but does not specifically interject proximate cause requirements into the strict products liability doctrine. See Restatement (Second) of Torts (1977), Sec. 402a.

35. 588 F. Supp., p. 428. Allen involved the effects of atomic radiation on human health, a trans-scientific issue similar to Agent Orange in which the relevant epidemiological studies are not complete. In *Allen* the district court shifted the burden of proof upon a showing that ionizing radiation was hazardous (based on animal studies and reports of human injuries resulting from occupational exposures and industrial accidents); that plaintiff was exposed to substantial concentrations of the radiation; and that plaintiff's injury was consistent with such radiation. In setting forth the rationale for this burden shifting, the court stated: "This shift in burden of proof reflects a sound application of important legal policies to the practical problems of trying a lawsuit: where a strong factual connection exists between defendant's conduct and the plaintiff's injury, but selection of 'actual' cause-in-fact from among several 'causes' is problematical, those difficulties of proof are shifted to the tortfeasor, the wrongdoer, in order to do substantial justice between the parties" (p. 411). See also S. 1480, 96th Cong. 2d sess., 1980, sec. 4; H.R. 1049, 96th Cong., 1st sess., 1979, sec. 101–106; H.R. 3797, 96th Cong., 1st sess., 1979, sec. 3211–15; H.R. 5291, 96th Cong., 1st sess., 1979, sec. 211–15; Senate Committee on Environment and Public Works, *Injuries and Damages from Hazardous Wastes—Analysis and Improvement of Legal Remedies: A Report to Congress in Compliance with Section 301(e) of the Comprehensive Environmental Response, Compensation, and Liability Act of 1980 (P.L. 96–510) by the Superfund Section 301(e) Study Group*. 97th Cong., 2d sess., 1982. S. Rept. 12. This is discussed at length by Strock in Chapter 9.

36. The term "unreasonably dangerous" should not be confused with the "abnormally

dangerous" terminology used in the Restatement (Second) of Torts for imposition of liability under strict liability theories. See sec. 520.

37. In fact, despite the no-fault scheme articulated under the strict liability doctrine, courts apparently consider the foreseeability of the risk in assessing liability. Consequently, this proposal would not change the case law but would only make the basis for judgments more explicit and predictable. See M. J. Polelle, "The Foreseeability Concept and Strict Products Liability: The Odd Couple of Tort Law," *Rutgers-Camden Law Journal* 8 (1976): 101, 103–109.

38. Traditional negligence differs from the unreasonably dangerous standard proposed here only insofar as "reasonableness" is concerned. Whereas the negligence standard holds defendant to the level of a "reasonable man," the "unreasonably dangerous" standard holds manufacturers to the level of an expert—responsible for all information available when products are marketed. Whether a particular manufacturer is aware of the hazards is immaterial since expert knowledge is imputed to every manufacturer.

39. The standard is based on all relevant information; that is, the supplier is presumed to know all the dangers that exist in a product at the time it is marketed.

40. Factors to be considered include the normal expectations of the consumer as to the manner in which the product will be used, the complexity of the procedures consumers must follow to use it, the magnitude of the danger to which the user will be exposed, and the likelihood of harm to the user. See J. Sales, "The Duty to Warn and Instruct for Safe Use in Strict Tort Liability," *St. Mary's Law Journal* 13 (1982): 521, 527. Also, the court in *Feldman v. Lederle Laboratories*, 97 N.J. 429, 479 A.2d 374 (1984), an action taken against a manufacturer of the drug tetracycline due to the side effect of tooth discoloration, undertook a similar analysis of "unreasonably dangerous," camouflaged as a discussion of strict liability: "[O]nce the defendant's knowledge of the defect is imputed, strict liability analysis becomes almost identical to negligence analysis in its focus on reasonableness of the defendant's conduct. [The standard is] would a person of reasonable intelligence or of the superior expertise of the defendant charged with such knowledge conclude that defendant should have alerted the consuming public?" (pp. 451–52, 479 A.2d, p. 385–86).

41. See *Osterdorf v. Brewer*, 51 Ill. App. 3d 1009, 1013–14, 367 N.E.2d 214, 217 (1977), where plaintiff could not recover damages resulting from injury when a gasoline cap flew off a fuel tank of a tractor because plaintiff admitted knowing of the necessity of tightly securing the gasoline cap after refueling.

42. This proposed standard does not address one set of cases—those in which plaintiffs were injured by a hazardous substance but the manufacturer cannot be held negligent because detection of the hazardous nature of the product was beyond the capabilities of contemporary science. Though the victims deserve compensation, imposing liability on the manufacturer would obviously not further deterrence. When a strict liability theory would not provide compensation, the legislature may wish to intervene and provide a compensation fund for the victims of certain unforeseen risks which society chooses to incur in pursuit of technical innovations and scientific progress. For a brief discussion of such compensation funds, see J. Trauberman, "Compensating Victims of Toxic Substances: An Analysis of Existing Federal Statutes," *Harvard Environmental Law Review* 7 (1981): 1, 3–4.

# 7

# Adapting Workers' Compensation to the Special Problems of Occupational Disease

## LAURENCE LOCKE

Workers' compensation systems have failed to meet the needs of the growing number of occupational disease victims. In the past half-century, we have learned a great deal about the hazards that complex chemical agents pose to those who must work with and around them. Yet workers' compensation statutes have not evolved accordingly. Most of them provide for periodic payments and medical care for wage earners suffering wage loss from injuries, diseases, or deaths that "arise out of and in the course of their employment."[1] Vestiges of a simpler time, they are designed to serve victims of acute, traumatic injuries occurring on the job. For a person suffering from a long-latent occupational disease, workers' compensation offers little hope of financial relief.

This chapter argues for the reformation of workers' compensation to address the unique problems posed by occupational disease that are discussed at some length in earlier chapters. Two premises underlie this argument. The first premise is that workers' compensation properly ought to be the primary source of support for occupational disease victims—those people suffering from diseases as a result of chemical exposures while on the job. An efficient workers' compensation system spreads among consumers the cost of work-related injury and disease occurring in the course of the production of goods and services. It creates an incentive for employers to maintain safer workplaces. And it grants each employer immunity from negligence liability in return for assuming the primary obligation to compensate occupational disease victims.

---

Adapted from the Harvard Environmental Law Review 9:2. © 1985 by the Harvard Environmental Law Review, President and Fellows of Harvard College.

Many would argue that adapting workers' compensation along the lines proposed here would be prohibitively expensive. There is no doubt that the suggested changes would entail substantial costs. In 1978, less than a decade after the National Commission on State Workmen's Compensation Laws suggested major improvements in state compensation systems, the annual cost of workers' compensation had more than tripled to $15.8 billion.[2] Meanwhile, the average employer premium rose by more than 60 percent and the incurred loss ratio for private carrier climbed from 63 to 78 percent.[3] One authority estimates that "(1) if just two percent of all cancer fatalities were determined to be job-related, the number of workers' compensation death awards would double; and (2) if only one percent of all cardiovascular fatalities were found to be occupationally caused, the number of death claims awarded annually would triple."[4] These additional death awards would cost an enormous amount of money. Even today, when permanent disability or death awards are made in but a small fraction of the total number of workers' compensation cases, they represent approximately 75 percent of all benefit payments.[5]

But emphasizing the costs of the proposed changes in the workers' compensation system tends to obscure the crucial issues. As one expert has observed, the question is not whether society can afford to pay for the costs of occupational diseases, but rather *who* should pay for them.[6] Although we may choose not to pay these costs in the form of workers' compensation, we may not avoid them altogether. At present, society at large foots most of the bill—through Social Security and welfare. In 1980, the annual cost of occupational disease absorbed by the Social Security and welfare systems was over $2.2 billion.[7]

This arrangement is inequitable and inefficient. It has been estimated that a mere 5 percent of severely disabled occupational disease victims are awarded workers' compensation.[8] If so, then as much as 95 percent of the burden rests on all of us, in the form of taxes, insurance premiums, and the like. Inefficiencies result because employers using hazardous chemicals are not forced to internalize their costs.

As is discussed by Billauer and Katzman in Chapters 5 and 8 respectively, alternative means of compensation—tort litigation, insurance and Social Security—are inherently ill-suited for remedying the special problems of the chemically exposed occupational disease victims. Small-scale employers often lack comprehensive group insurance and disability programs. Where such programs exist, benefits may be unevenly and sporadically distributed. Social Security beneficiaries, for example, are subject to strict eligibility requirements. According to the U.S. Department of Labor, more than 50 percent of occupational disease victims are declared ineligible for Social Security benefits.[9] Furthermore, Social Security, which requires a beneficiary to be totally disabled, provides no coverage to those suffering from permanent partial disabilities.[10] Recovery through tort litigation by an action against a third party raises the same set of problems that workers' compensation laws were designed to alleviate: difficulties of proving causation, judgment-proof defendants, and long-delayed recoveries.

Hence, the first premise: workers' compensation should serve as the primary source of support for workers harmed by hazardous chemicals. A well-crafted workers' compensation system would effect a redistribution; it would require employers (and their customers) to assume responsibility for occupational diseases. Despite the apparent expense of reforming it, the workers' compensation system is the best means of compensating victims of work-related chemical contamination.

The second premise is that legislative reform of workers' compensation is better pursued at the state than at the federal level. The inadequacies of state statutes have led to calls for federal benefit programs, as discussed at some length by Strock in Chapter 9. Such programs would extend employer immunity by barring product liability and other third-party suits. Costs would be transferred from individual employers to entire industries and to taxpayers. These schemes promise prompt payment with minimal controversy, although they entail establishing a ponderous claim-processing bureaucracy. But they are bitterly contested and may be years away from enactment. In the meantime, inequity and injustice persist. In light of the urgency of the problem and the uncertainty of federal intervention, the best practical solution is to work for reform of the existing state acts.

The next section of this chapter provides a historical background of workers' compensation laws and describes early attempts to compensate workers stricken by occupational diseases.[11] Recognizing the difficulties that contemporary victims of chemical contamination face in attempting to qualify for workers' compensation, it posits some explanations for underutilization of the system. The final section lists specific proposals for revising workers' compensation laws and simplifying their administration.

## HISTORICAL BACKGROUND

### Overview

The traditional view is that workers' compensation arose out of a dissatisfaction with the remedies available for work-related injuries at common law and under state employers' liability acts.[12] However, some scholars believe that the growing number of successful tort suits by injured workers against the employers—and not humanitarian zeal—precipitated the flurry of workers' compensation legislation. One study found that between 1875 and 1905 plaintiffs recovered damages in employer negligence suits more frequently than ever before.[13] Another authority has observed that fear of a trend of "substantial damages awarded by sympathetic juries" marked the period preceding the passage of workers' compensation statutes.[14] Whatever the correct explanation, a great many states first enacted workers' compensation statutes in the decade between 1910 and 1920.[15] The new laws established a no-fault compensation system for work-related injuries that served as the workers' exclusive remedy.[16] Employers thus were granted immunity from tort liability in exchange for providing a minimum level of protection to their employees.[17]

Most earlier workers' compensation laws did not expressly apply to disease.[18] Some legislators, uncertain as to the appropriate scope of the new laws, intentionally bypassed the issue of occupational disease, while others preferred to leave the matter for the courts. Legislators were primarily concerned with traumatic injuries, such as machinery-related accidents, where there was no doubt that the employee was harmed while on the job. By broad judicial construction of the phrase "personal injury," Massachusetts became the first state to extend coverage to what we now think of as an occupational disease. Other states were slow to follow suit. But the growing success of plaintiffs in common law actions against employers for work-induced diseases not covered by compensation acts—and thus not barred by exclusive remedy clauses—led to a drive to bring these diseases within the domain of the new acts.[19] Some jurisdictions passed separate occupational disease laws that paralleled workers' compensation acts in providing limited relief for employees and employer immunity.[20]

## Workers' Compensation Statutes and Occupational Disease

*Occupational Disease Coverage.* Historically, provisions for coverage of occupational disease could be put in three general categories. Laws in the first category featured a schedule of diseases identified by the nature of the disease and the occupation of the worker.[21] These laws had the virtue of simplicity; where they applied, they were a form of prima facie coverage, and workers faced minimal problems of proof. The theory promoting the adoption of schedules was that insurers would be inclined to pay benefits without contesting worker eligibility once a prima facie showing of injury had been made.[22] Problems arose, however, when diseases not listed in the schedules afflicted workers. Only the legislatures could amend the schedules to include more diseases. A sufficient number of workers had to suffer from a condition before the medical profession deemed it an occupational disease, and then the legislatures had to be persuaded to add the disease to the schedules. Meanwhile, workers suffered or died without compensation.

The second general type of occupational disease provision contained a schedule followed by a catch-all clause. In this scheme the schedules did not purport to be exhaustive; the residual clauses sometimes covered unscheduled diseases. These laws had the advantage of the presumption offered by the schedules and catch-all clauses that allowed for greater liberality and flexibility in awarding benefits.[23]

Laws in the third category eschewed the schedule approach and instead set forth general definitions of occupational disease. These definitions were included either in the existing workers' compensation acts or in separate occupational disease laws. The usual definitions limited covered occupational diseases to those "peculiar to" or "characteristic of" the relevant trade. Other definitions excluded "ordinary diseases of life" or required a "hazard greater than that to which the general public was exposed." Still others required that the employment be the "proximate cause" of the disease.[24] All of these laws attempted to distinguish "occupational" and "ordinary" diseases by means of criteria that had no medical significance and were difficult to apply.

*Special Restrictions.* Many occupational disease laws imposed other restrictions on coverage of occupational diseases. Minimum exposure times were required for certain diseases. Unless an employee had worked for a specified period for one or more employers, he was ineligible for benefits.[25] Similar provisions disqualified those who had not worked for an in-state employer for a sufficient time period.[26] For example, a coal miner in Kentucky developed black lung disease and applied to the state workers' compensation commission for total permanent disability relief. The commission denied his petition because Kentucky law required that the employee be exposed to the hazard continuously for two years prior to the illness. In the two years before his application, the miner had been absent from work for four months due to a back injury he had suffered while at work. The commission ruled that, notwithstanding the employee's blamelessness, his absence disqualified him because of the minimum continuous exposure requirement. A reviewing court upheld the commission's decision on appeal. Apart from being medically irrelevant, these minimum exposure provisions militated against workers in certain trades, such as the construction and insulation industries, in which employees habitually move from job to job or state to state.[27]

A further set of provisions established special requirements of proof for occupational disease victims. Special medical panels of industrial disease referees reviewed all occupational disease claims. The referees' opinions either were binding on compensation administrators or were of great persuasive force—even though the referees were not subject to cross-examination.[28]

*Inadequacy of Provisions Designed for Traumatic Injury Claims.* Provisions better suited for traumatic injuries than for occupational disease pervaded early legislation. Statutes of limitation that were reasonable as applied to traumatic injury cases imposed harsh results when applied to occupational disease victims.[29] These restrictions barred claims for many diseases that had long periods of latency.

Moreover, workers' compensation acts commonly set weekly compensation benefits as of the date of injury.[30] This practice resulted in almost ludicrous injustices in long-latent disease cases. Just before World War II, in Salem and Lynn, Massachusetts, the fluorescent light industry used a fluorescing powder containing beryllium oxide. The industry stopped using beryllium (which was needed for the atomic bomb) by the end of 1942. Several hundred young women who had worked in the industry's factories contracted a disease from exposure to beryllium oxide. Some workers began coming down with the disease as early as six months after exposure. However, many showed no discernible symptoms for as long as thirty years. Nonetheless, their compensation (or that of their spouses if they were married when exposed) was limited to two-thirds of the wages they had earned in the early 1940s—$15 to $22 per week—with a maximum in death cases of $10 weekly.[31]

*Employer Reaction.* The impact of even the limited coverage afforded by early laws brought threats of retaliation from employers, who warned that they would have to shut down their plants because of the potential economic impact of the new laws.[32] These threats persuaded even some unions to join in the drive to limit available compensation. Legislatures reduced benefit levels by eliminating partial incapacity compensation, reducing the maximum period of payments, and requiring that death ensue within a prescribed period.[33] The net effect of the changes was to allocate to workers most of the risk of work-related illness. Workers were forced to accept this risk in order to retain their jobs.

## Contemporary Problems

Since 1970, the American public has woken up to the dangers of toxins in the workplace. This development has led to improvements in state and federal workers' compensation acts. After the National Commission on State Workmen's Compensation Laws issued its report in 1972, state legislatures, encouraged by employers and insurers who feared federalization of the state-run compensation systems, amended their laws to provide for more general coverage of occupational disease.[34] Nonetheless, the main features described above still pervade workers' compensation laws. An anachronistic legal framework remains in place; it has not met the tremendous challenges presented by chemical contamination in the workplace.

*Litigational Obstacles.* The litmus test of a successful workers' compensation act is the extent to which insurers voluntarily honor claims. In an ideal system, the administering agency must resolve disputes in only a small percentage of cases. With respect to occupational disease claims, workers' compensation fails this test. In these cases, insurers rarely pay voluntarily. According to a 1975 survey, 62.7 percent of all compensated occupational disease claims were controverted. Causation was the dominant issue in 72 percent of those cases. Insurers challenged 88 percent of the dust disease claims and 78 percent of the claims involving respiratory diseases due to toxic agents. By contrast, fewer than 10 percent of the accident claims raised any dispute.[35] Insurers may be loath to agree that a claimant's disease stems from exposures to chemicals used in the workplace or to concede that the claimant's disability or death is a product of the disease. It has been my experience that since usually no penalty for delay exists other than interest assessments, insurers withhold payments and hope that desperate claimants will accept relatively small settlements.

The full range of litigational obstacles, and the various means by which to bridge them are discussed at length by Wagner in Chapter 6. Many of these obstacles have been circumvented for asbestos damages, the class of occupational disease for which the litigation experience has been the most extensive. It pays here to compare the findings on asbestos to average settlement statistics for all occupational disease claims, of which asbestos is but one.

The average worker suffering from occupational disease in the late 1970s received only a 40 percent replacement of lost wages and had to wait one year for the first payment.[36] In more than 60 percent of occupational disease cases, the employer controverted the claim.[37] Only 15 percent of challenged claimants won their cases.[38] Those fortunate enough to prevail received little in compensation. The total compensation ranged from a minimum mean of $1,174 (for temporary partial disability) to a maximum mean of $9,676 (for permanent total disability).[39]

The average settlement for asbestos cases is about $50,000, with between 11 and 65 percent (depending on the state) disposed of within five years. Some settlement may come well before final disposition of the case, but less than 30 percent of the cash is received at that time. Most asbestos workers bypass the workers' compensation system, and about half win their claims.[40]

In light of these findings, it is no surprise that a mere three cases out of every one hundred incidents of occupational disease lead to workers' compensation claims.[41]

*Underutilization.* As we have seen, most occupationally diseased workers decide not to file claims. In 1980, for example, only sixteen workers' compensation claims for work-related cancer were filed in Ohio.[42] The records of some state commissions in awarding benefits are so poor, and the litigational and financial obstacles to recovery so formidable, that many workers seek alternative means of compensation.[43] A 1974 survey reported that 53 percent of those afflicted with occupational diseases relied on Social Security as their major source of income support, compared to only 5 percent relying on workers' compensation.[44] Even in the area of asbestos-related diseases, only about 15 percent of all disabled asbestos workers received workers' compensation benefits prior to their deaths.[45]

Additional factors contribute to the underutilization of the workers' compensation system. Many workers—and their physicians—are unaware that they have a condition for which compensation might be available.[46] This ignorance persists despite the publicity given to occupational disease in the press. If a victim never associates his disease with his occupation, it follows that he will not bring a claim or consult an attorney.

Moreover, even where a worker knows that he has a compensable occupational disease, he may be affected by fear. An ill, discouraged worker may prefer to abandon his rights rather than to begin an expensive, arduous lawsuit. Many workers succumb to fatalism, believing that they have no choice but to accept their misfortune.

A final cause of the underutilization of workers' compensation, particularly among more affluent workers, is the availability of other resources. Many employers have group medical insurance and short-term and long-term disability programs. These first-party programs pay benefits without proof of work-relatedness where the employee has medical bills or a disability. Many workers have recourse to Social Security, which compensates totally disabled workers, pays retirement benefits to those over sixty-five, and supports a worker's survivors if he dies. A 1974 survey revealed that Social Security is the most common source of income for the totally disabled.[47] In addition, for some workers there is the possibility of a third-party tort suit. In such a suit, a worker seeks tort damages outside the workers' compensation system. For example, an injured employee might sue for negligence a manufacturer of a toxin that his employer uses in the workplace. Many of the asbestos cases include this type of suit. Because workers' compensation insurers must be reimbursed from the proceeds of a third-party settlement or judgment, many claimants and their attorneys see no reason to bring both a compensation case and a third-party suit.[48]

For all of these reasons—practical and financial obstacles in litigation, ignorance, fear, fatalism, and the availability of other resources—the workers' compensation system is today severely underutilized. There can be no doubt that the system has failed miserably to redress the injustices suffered by occupational disease victims.

## ADAPTATION OF WORKERS' COMPENSATION

The frequent necessity of litigation, the financial and practical barriers to access to the system, and the inherent incongruities of scale between victims and insurers are generic deficiencies that affect all workers' compensation claimants. In addition to these systematic flaws, the workers' compensation system is subject to the same barriers to compensation that the tort litigant must face: causation, multiple causation, latency, statutes of limitations, claim preclusion, inflation, and partial disability. Because Billauer and Wagner have discussed these at some length, no further treatment will be given in this chapter. Nonetheless, these problems—even apart from the system's more general deficiencies—are sufficient to justify an extensive revision of the workers' compensation laws.

### Specific Proposals for Revising Workers' Compensation

The area of toxic exposure in the work environment is both complex and expanding. The problems of identification, multiple causation, latency, and others unique to chemical contamination call for a new approach to the compensation of occupational disease victims. We can now identify specific means of adapting workers' compensation laws to serve these victims more effectively.

*Inclusive Definitions.* All definitional limitations on compensability should be removed from the state statutes.[49] Statutes should be construed to include any disease that arises out of and in the course of employment. Although critics may contend that eliminating definitional limitations would in effect convert workers' compensation into national health insurance, the continued requirement of proof that a compensable disease results from a condition or feature of employment would prevent such a conversion. A broad definition would eliminate most of the grounds for controversy that now plague the administration of these acts. Borderline cases should be left for case-by-case adjudication.

*Aggravation of Preexisting Conditions.* No claimant should be denied compensation on the ground that workplace exposure to toxins has aggravated a preexisting condition. No distinction should be made between a preexisting condition that has its origin in a prior work injury or exposure and conditions caused by non-work-related illnesses or personal predispositions such as alcoholism and cigarette smoking. No attempt should be made to measure the degree by which work-related contamination contributes to the claimant's ultimate condition, so long as the impact of the exposure is measurable.

*No apportionment.* The doctrine of apportionment should be abandoned in chemical contamination cases. Apportionment would require that an employer compensate a diseased employee only to the extent that work-related factors have caused the claimant's condition. If the claimant has been exposed to toxins at a series of different jobs, apportionment requires that the contribution of each exposure be measured and that compensation be paid by each employer in proportion to its share.

These tasks are inordinately difficult. They increase the dependence of both claimants and defendants on medical experts who profess to disentangle the inextricable. They encourage insurers to controvert claims. Simplicity requires that claimants, insurers, and tribunals not be forced to separate the effects of workplace contamination and other factors.

*"Last Exposure" Rule.* The rule now in use requires the last employer at whose workplace the claimant was demonstrably harmed by a toxin to bear the full burden of compensation.[50] This is called the "last injurious exposure" rule. The rule has the virtues of medical logic and apparent fairness. Arguably, an employer should not be liable to a worker who, though exposed to a toxin at the employer's workplace, had been harmed irreparably by prior exposure at a different workplace. Yet because of its emphasis on determining injuriousness, the rule is fraught with all the difficulties of apportionment; it obliges the claimant to bring in all the employers for whom he worked where there was even a possibility of harmful exposure. As to a disease of long latency, it may be impossible for medical experts to determine which exposure filled the disease cup to the brim and thus rendered immaterial all later exposures. In practice, in such cases the claimant's attorney and the insurers agree upon an appropriate level of compensation and all the relevant employers contribute on a pro rata basis. The effect is de facto apportionment.

A better rule would require the last employer at whose workplace the claimant was exposed to the relevant toxin—regardless of whether the exposure was additionally harmful or not—to compensate the employee. The choice between the "last exposure" and "last *injurious* exposure" rule should be guided by both practical considerations and a recognition of the underlying purposes of compensation. The goal should be to promote simplicity and ease of application, so as to encourage voluntary payment of compensation by insurers and to facilitate adjudication in the event of controversy. The last exposure rule, by removing an element of uncertainty—injuriousness—would further these ends. But would it cause actuarial unfairness? In many cases an employer might be required to compensate a worker whose condition was not proximately caused during the worker's tenure with the employer. Insurers, however, can plan ahead for rate making under the last exposure rule and can create reinsurance pools to spread their losses. This sort of flexibility would be an important component of a comprehensive legislative reform package.

*Minimum Time Periods.* Restrictive time limits that specify minimum periods of employment or in-state work should be abandoned. These artificial and ill-conceived provisions unjustifiably hinder occupational disease victims seeking compensation. They discriminate arbitrarily and have no medical justification. Their restrictive effect is at odds with the very purpose of workers' compensation: to compensate equitably those injured or harmed on the job.

*Liberalized Statutes of Limitation.* A limitation period running from the date of exposure is clearly inappropriate in occupational disease cases, which may feature long latency periods.[51] Today many cases are barred by statutes of limitation because work relatedness was not diagnosed at an early stage. Some advocate adoption of the judicially created tort rule that the period of limitation runs from the "date of discovery," meaning the date on which the plaintiff knew or should have known that he had sustained an injury or disease for which the defendant would be liable.[52]

Although a step in the right direction, this rule might be unduly restrictive in chemical contamination cases, especially when the disease has a long latency period. In a long-latent disease case, the diagnosis may precede the first signs of work incapacity. For several reasons a diseased worker may not want to file a claim upon first discovering his illness. First, the employee may initially incur only minor medical expenses, which may be fully covered under a group health insurance policy. Second, a stricken employee may be able to work effectively for years after the discovery of his illness. He may be transferred to a less hazardous or demanding job, or go to work for a different employer. Third, a worker may not deem it worth the requisite effort involved in establishing a claim, particularly when the claim may not yet be compensable. A workers' compensation commission typically will not act on a claim until sufficient evidence of disability exists to warrant a hearing. In short, the date of discovery rule may force a diseased worker to file a claim prematurely.

A better rule would protect the employee with valid reasons for delaying his claim. The applicable statute of limitations should run from the date that the claimant first became totally incapacitated and knew or should have known that his disability was work related. This rule would eliminate the requirement that a diseased worker prematurely file a claim and would preserve the worker's right to compensation when the need arises.

*Adjusted Benefit Rates.* Compensation levels should be based on the benefit rates in effect on the date of disability or death, rather than on the date of exposure. The injustice in basing benefits on date of exposure rates is intuitively apparent. However, if this proposal was adopted for all pending cases, it might be necessary to create an insurance pool for retroactive claims to reimburse each insurer for the difference between the benefit rate on the date of exposure, when the premiums were set, and the rate on the date of loss. For future cases, insurers could take into account the effects of the new rule in setting rates.

*Modified Benefit Structures.* Benefit structures must be modified to reflect the special character of occupational disease disability. Setting partial incapacity compensation by comparing present earning capacity with a fixed preinjury wage fails to take into account the escalation of wage levels. Where partial incapacity is demonstrated, benefits should be based on *loss of earning capacity*. The employee's impaired earning capacity should be compared with what his earning capacity would have been had it not been impaired. This calculation requires two figures. The first figure is found by extrapolating from the employee's wage at the time of last exposure—the date of injury. This projection would be based on an analysis of wage-level increases over the relevant time period, which may be as long as twenty years. The second figure is simply the wage that the impaired claimant is presently able to earn. The difference between the two figures determines the rate of compensation for a partially disabled employee. If the claimant becomes totally incapacitated or dies, he or his survivors should receive compensation at the rates then in effect.

In some cases an occupational disease victim may suffer in ways that do not affect his earning capacity, such as loss of hearing or reproductive functioning. In these, compensation commissions should employ an additional benefit schedule. Appropriate adaptation of workers' compensation requires the availability of benefits both for loss of earning capacity and for other physical impairments.

## The Problem of Proof

The difficulties involved in proving occupational disease claims plague the workers' compensation system. These difficulties affect insurers' decisions as to whether to accept claims. They affect compensation administrators, who must decide controverted cases. Most of all they affect claimants, who must succeed in convincing lawyers to represent them in what are often time-consuming, difficult, and expensive cases. Problems of proof also bear upon the choice the lawyer and her client must make among the available alternatives: workers' compensation, and other income maintenance systems such as long-term disability and Social Security, and tort litigation against third-party defendants. Because of the centrality of problems of proof in occupational disease cases, ways of reducing their significance must be found.

*Proof of Chemical Contamination in the Workplace.* Employers must be required to keep records of the toxic substances used in their operations and the locations at which such substances are used. The collection and maintenance of these records should be mandated, as they have been in some states.[53] A centralized computer bank should be established for storage of the data. Insurers and claimants should have access to the data for evidence of workplace exposure.

For certain highly toxic substances, workplace monitoring should be required. In some situations, such as those involving radiation, individual employees should undergo regular exposure-level tests.[54] Monitoring data should be collected and maintained in a central computer bank, with access available to insurers and employees alike.

The creation of such a system of workplace monitoring and data collection would require a commitment to place the needs of the victims of toxic hazards above the so-called right of employers to the privacy of their work processes. It would also require a significant commitment of societal resources.

*Proof of Employee Contamination.* Legislation should require employers to provide periodic examinations for their employees whose work exposes them to toxins. A limited number of industries are now required to fund such periodic examinations. In other instances, worker health surveys have been performed by unions and health centers under OSHA grants.

One problem with employer-initiated screening is the failure of personnel departments to notify workers of pathologic conditions discovered in the course of examinations. A worker may be denied information essential to an informed decision as to whether to continue working in a contaminated environment, as happened to asbestos workers in the late 1950s.[55] Another difficulty is the panic that may strike an employer who discovers a health problem. Such an employer may hastily terminate long-term employees. A third problem is the difficulty that already exposed job seekers have in finding employment, especially if a new employer may be held responsible for compensation under the last exposure rule.[56] The fourth difficulty is that screened employees can cause fear and hypochondria to spread through their workplaces. Despite these problems, however, periodic screening of exposed workers is an essential preventive and diagnostic procedure and would prove invaluable in contested occupational disease cases.

Physicians must take steps to increase their knowledge of occupational diseases, as well. Accurate medical diagnoses of chemically induced illness and death are particularly important to an effective workers' compensation system. Diagnoses must be informed by up-to-date medical knowledge.

The information gap is slowly narrowing, as we can see from Schnare's discussion in Chapter 2. Increasing numbers of physicians now specialize in occupational medicine, and growth in the diversity of diagnostics and patient care, as presented by Root and Schnare in Chapter 4, indicates that some degree of hope is on the way. However, physicians who provide primary care for persons exposed to chemical hazards must learn to take detailed occupational histories and must avail themselves of opportunities that will expand their knowledge base and expertise.

An attorney or physician interested in a worker's previous exposures to toxins should also be able to undertake a comprehensive review of his work history—

and should do so. In long-latent disease cases in which employees have held many jobs in different parts of the country, it is useful to resort to the Social Security Administration, which maintains an earnings record for each employee.[57] This record often refreshes the claimant's recollection and helps to establish instances of exposure. The worker, perhaps with the assistance of his union, should also be expected to maintain a lifelong record of work-related chemical exposures, and midcareer workers should be encouraged to create such a log and keep it current.

*Problems of Medical Proof.* A claimant in an occupational disease case, just as in a traumatic injury case, cannot prevail unless he shows that his disease was caused by exposure to toxins. The full breadth of this problem is presented in Chapter 5, and it deserves a full reading. This problem is more theoretical, however, than it is real. In practice, a medical expert will give his opinion as to the existence of a causal relation. He will be asked to state the bases for his opinion. These will include the claimant's history, tests, and medical records, and the witness's own expert knowledge. This expert knowledge may include the results of epidemiological studies. Thus, such studies may properly serve as a substratum underlying the expert's opinion, but the tribunal need not make a finding based on the studies alone.[58] On cross-examination, the expert can be questioned as to the degree to which his opinion is based on epidemiological studies. If he insists that the studies form but one of several grounds for his opinion and that his opinion is not speculation, his evidence will support an award.

In controverted occupational disease cases, the medical experts called by the two sides often disagree as to causality and degree of impairment. Several critical studies decry these variances. Apparently they assume that two truly impartial experts, relying solely on studies of the claimant and the received medical wisdom, would not so often provide disparate opinions. Conflicting expert opinions are said to increase the likelihood that insurers will not voluntarily accept occupational disease claims. The studies also suggest that disparate opinions merely confuse the workers' compensation hearing officer, who must rely on instinct in lieu of settled scientific opinion; the resulting system resembles a lottery.[59]

This complaint is not unique to the occupational disease arena. Contradictory expert opinions often arise in product liability cases, in personal injury suits, and in criminal trials in which forensic psychiatrists and pathologists testify. They are a common ingredient in all litigation. The opportunity of adverse parties to call on expert witnesses to substantiate their respective contentions is an essential element of due process. It is just as vital to our system of justice as the right to have a trial judge, jury, or compensation hearing officer sift through the facts and decide the case on its merits.

Nonetheless, some commentators have called for the prohibition of expert witnesses in occupational disease cases. They propose to transfer the authority to decide medical issues to a neutral body of experts. These experts would apply their knowledge to the problems of medical proof in toxic disease cases, and in so doing—so the theory goes—introduce certainty where nought but confusion now exists. The opinions of such bodies would be accorded considerable deference.[60]

Despite the initial attractiveness of these ideas, they are subject to cogent criticism. If they were implemented, adverse parties would have no opportunity to test the impartiality and medical competence of such panels or to impeach the bases of their opinions. It must be recognized that any expert panel will consist of individuals, with their own biases and limitations.

These objections to obligatory medical panels apply with even greater force to proposals for transferring the entire decision-making authority to expert claims-examining units. Such units would both decide medical issues and rule on the merits of individual claims. It would be hard to imagine a scheme more at odds with the tenets of our adversary system.

Wide disparities in medical opinions do not necessarily reflect the mercenary biases of expert witnesses. Some disagreement must be expected in the unsettled toxic contamination field. The solution is not to get rid of the experts, but to reduce the degree of uncertainty. We need further medical and scientific investigations of toxic hazards and their impacts on human beings. Justice will be better served by an increase in our knowledge of occupational disease than by the establishment of all-powerful expert bodies to decide individual cases.

## Coordinating Workers' Compensation and Tort Litigation

The interplay between tort recovery and workers' compensation benefits constitutes the final issue on our agenda of reforms. Where a third party—such as a manufacturer of toxins later used in an employer's workplace—is responsible for causing a worker's disease or death, it should be held liable for the resulting damages. Yet we have maintained that workers' compensation should play a fundamental role in providing support for occupational disease victims. How ought we to reconcile these potentially conflicting goals? In some cases workers who have already qualified for and received workers' compensation benefits bring successful tort suits against negligent third parties. Most workers' compensation acts require that in the event of recovery from a third party, the workers must reimburse the compensation insurer for the remaining sum. Under this arrangement, the workers' compensation insurer always is reimbursed. The effect is to relieve the workers' compensation system of its fundamental responsibility to compensate injured workers merely because of the fortuity of the third party's negligence.

In an attempt to address this situation, some have proposed that the tort recovery be reduced to the extent of any amount received through workers' compensation. For example, if an employee had received $50,000 in workers' compensation and then won a tort judgment of $100,000 against a third party, he would receive only $50,000 from the third party. This would reward the wrongdoer—the negligent third party—and in some cases relieve it entirely of its obligation to pay damages.

A compromise rule would preserve the primary role for workers' compensation while ensuring that negligent third parties are held fully responsible for their actions. The rule would require a successful tort plaintiff to reimburse the workers' compensation insurer for one-half the amount paid in compensation. For example, if an employee had received $50,000 in workers' compensation and won a tort judgment of $100,000, he would be required to pay back $25,000 to the workers' compensation insurer. Although theoretically it would permit an employee to recover a windfall, this approach would advance both of our stated goals while avoiding the pitfalls involved in apportionment.

## CONCLUSION

Workers' compensation has failed to provide significant relief to more than a small percentage of occupational disease victims. The explanation of this phenomenon lies in part in systematic deficiencies and in part in special problems faced only by claimants in occupational disease and toxic chemical exposure cases. If the workers' compensation system is to fulfill its mission as the fundamental source of support for workers disabled on the job, we must make a number of statutory adaptations. In addition, we must introduce a variety of measures that diminish the complex problems of proof that hinder claimants harmed by chemical contamination. The proposals advocated in this chapter will help to ensure that the injustices suffered by the victims of toxic chemical exposure do not go unredressed.

## NOTES

This chapter was first published in the *Harvard Environmental Law Review*, 9, no. 2, 1985: 249–82, and has been adapted for use in this volume with permission of the President and Fellows of Harvard University.

1. These words were borrowed from the 1897 English Workers' Compensation Act and have been adopted in some forty-two states. See generally A. Larson, *The Law of Workmen's Compensation* (New York: Matthew Bender, 1984), sec. 6.10. They have been characterized as "deceptively simple and litigiously prolific." *Cardillo v. Liberty Mut. Ins. Co.*, 330 U.S. 469, 479 (1947).

2. See note 11. See also, Assistant Secretary for Policy Evaluation and Research, U.S. Department of Labor. *An Interim Report to Congress on Occupational Disease* (1980), p. 66 [hereinafter cited as 1980 Report].

3. 1980 Report. In addition, the incurred loss ratio is the relationship between premiums earned and losses incurred (or benefits paid). Until about fifteen years ago, this ratio remained stable at 60 percent. That is, sixty cents of every premium dollar was dispersed to employees in the form of cash or medical benefits. In 1970 the incurred loss ratio was 63.3 percent for private carriers. By 1976 it had risen to 78.7 percent. See L.

Darling-Hammond and T. Kniesner, *The Law and Economics of Workers' Compensation* (Santa Monica, Calif.: Rand Institute for Civil Justice, 1980), pp. 5–6.

4. *Proceedings of the Chief Executive Officers' Workers' Compensation Conference*, Airlie, Va., June 7–9, 1978, pp. 213–37 (remarks of Peter S. Barth).

5. 1980 Report, p. 66; and Darling-Hammond and Kniesner, *The Law and Economics*, pp. 4–5.

6. P. Barth and H. Hunt, *Workers' Compensation and Work-Related Illnesses* (Cambridge: MIT Press, 1982), p. 285.

7. 1980 Report, p. 4.

8. See note 51 and accompanying text.

9. 1980 Report, p. 81.

10. Ibid., p. 85.

11. Several significant studies of the workers' compensation system have appeared in the last seventeen years. The National Commission on State Workmen's Compensation Laws, established by the Occupational Safety and Health Act of 1970, issued a comprehensive report in 1972; *The Report of the National Commission on State Workmen's Compensation Laws*. The U.S. Department of Labor has conducted subsequent studies. See G. Weber and M. Lucero, *An Analysis of the Status and Adequacy of the Workers' Compensation System for Compensating Victims of Occupational Disease* (1980), a report prepared by the Health and Human Sciences Division of Teknekron Research, Inc. for the Office of the Assistant Secretary for Policy, Evaluation, and Research, U.S. Department of Labor, under contract. For a useful study with respect to occupational disease, see the report of Crum and Forster's Occupational Disease Task Force, *Role of the State Workers' Compensation System in Compensating Occupational Disease Victims* (June 1983) [hereinafter cited as Task Force]; See also Barth and Hunt, *Workers' Compensation*.

12. See, for example, L. Locke, *Massachusetts Workmen's Compensation Laws* (1981) Sec. 1; see also *LaClair v. Silberline Mfg. Co.*, 379 Mass. 21, 383 N.E.2d 867 (1979); *Young v. Duncan*, 218 Mass. 346, 106 N.E. 1 (1916). For a discussion of employers' liability acts, see Larson, *The Law of Workmen's Compensation*, 1, sec. 4.60; National Commission on State Workmen's Compensation Laws, *Compendium on Workmen's Compensation* (1973), pp. 13–14 [hereinafter cited as *Compendium*].

13. Posner, "A Theory of Negligence," 1 *Journal of Legal Studies* 29 (1972), cited in *Compendium*, p. 17.

14. Carl Gersuny, *Work Hazards and Industrial Conflict* (Hanover, N.H.: University Press of New England, 1981), p. 101.

15. The laws were usually enacted following studies by state commissions of precompensation systems. See Walter Dodd, *Administration of Workmen's Compensation* (London: Oxford University Press, 1936), pp. 29–38.

16. See generally Daniel Berman, *Death on the Job* (N.Y.: Monthly Review Press, 1978), pp. 19–20 (new workers' compensation systems were intended to "substitute a fixed, but limited charge for a variable, potentially ruinous one").

17. For a discussion of the exclusive remedy doctrine, see generally Larson, *The Law of Workmen's Compensation* (1983), 2A, subsec. 65.00–65.30.

18. 1980 Report, p. 67.

19. See Barth and Hunt, *Workers' Compensation*, pp. 2–3 (citing estimates that damage suits of an aggregate amount exceeding $300 million were initiated in the early 1930s).

20. Ibid. (for example, California and Wisconsin).

21. See generally Larson, *The Law of Workmen's Compensation*, 2, Sec. 57.14(c).

22. Ibid.

23. See Larson, *The Law of Workmen's Compensation*, 1B, sec. 41.10.

24. See Task Force, Appendix B (lists status of present state laws in defining occupational disease).

25. For the present status of statutory time restrictions on claims, see Barth and Hunt, *Workers' Compensation*, table 4.3, p. 122.

26. *Yocum v. Overstreet*, 512 S.W.2d 940 (Ky. Ct. App. 1974). The relevant law has been amended, *see* Ky. Rev. Stat. Sec. 342.316(5) (1984), but still yields some peculiar results. See, for example, *Yocum v. Eastern Coal Corp.*, 523 S.W.2d 882 (Ky. Ct. App. 1975). In *Eastern Coal*, an employee applied for compensation for black lung disease. Despite his thirty-six-year employment with Eastern Coal Corporation, he received only 75 percent of the benefits to which he otherwise would have been entitled because of a two-day stint with a different employer in 1948.

27. For an example of an in-state exposure requirement, see *State Compensation Fund v. Joe*, 25 Ariz. App. 361, 543 P.2d 790, (Ariz. Ct. App. 1975) (claim barred because last exposure occurred in Colorado instead of Arizona). The statute has since been amended, *see* Ariz. Rev. Stat. Ann Sec. 23–901.02 (1983), but the changes would not reverse the outcome of this case. See also Barth and Hunt, *Workers' Compensation*, p. 122 (table of state/time limitation rules).

28. Fully binding referees' opinions were in some cases constitutionally objectionable. See, for example, Meunier's Case, 319 Mass. 421, 66 N.E.2d 198 (1946). The provision, Mass. Gen. Laws, ch. 152, sec. 9B, was officially repealed in 1947 (1947 Mass. Acts 286).

29. Some statutes covered occupational "accidents." In applying statutes of limitation to these provisions, most courts held that the relevant time period began at the time of the specific work incident. This ruling effectively barred occupational disease victims from recovering benefits. They often had no disability or need for medical care until long after exposure to toxins. Only in Tennessee and Nebraska did this rule not obtain. See P. J. Kelley, "Statutes of Limitations on the Era of Compensation Systems: Workmen's Compensation Limitations Provisions for Accidental Injury Claims," 27 *Washington University Law Quarterly* 541, 564–66 (1974).

Note that other statutes applied to occupational "injuries." Occupational disease claimants fared better under these provisions. Although eight of the states with "injury" statutes interpreted the term to be equivalent to "accident," fifteen states were more flexible. In these states, the statutes of limitations began to run at one of the following times: (1) the time the injury became manifest, (2) the time the injury became compensable, or (3) the time the employee knew or should have known the compensable nature of his injury.

30. We have not yet eradicated this injustice. In 1977 the widow of a New Jersey asbestosis victim received benefits of $15 per week because her husband's wages at the time of his last employment with an asbestos company, 1922–24, were $17.75 per week. When he died nearly a half a century later, the employee was earning more than $400 per week. See Asbestos Compensation Coalition, *Compensating Workers for Asbestos-Related Disease* (1982), p. 9.

31. L. Locke, *Massachusetts Workmen's Compensation Laws*, sec. 195. For details on the plight of workers inflicted with beryllium disease, see Barth and Hunt, *Workers' Compensation*, p. 128.

32. See generally Berman, *Death on the Job*, p. 27.

33. In Massachusetts, the granite cutters' union successfully lobbied for the passage

of restrictive laws governing silicosis and other occupational dust diseases. See 1950 Mass. Acts 220 (repealing Mass. Gen. Laws ch. 152, subsec. 76–85). See also *Compendium*, pp. 16–17.

34. All fifty states now have laws providing workers' compensation for occupational disease victims. In addition, all the states except Florida, Mississippi, and Tennessee have adopted at least ten of the nineteen specific proposals set forth in the 1980 Report.

35. Barth and Hunt, *Workers' Compensation*, tables 5.16, 5.18, and 5.34.

36. 1980 Report, pp. 2, 59–60.

37. Barth and Hunt, *Workers' Compensation*, table 5.16, p. 163.

38. Ibid., table 5.19, p. 165.

39. Ibid., table 5.31, p. 174.

40. D. R. Hensler, et al., *Asbestos in the Courts: The Challenge of Mass Toxic Torts*, Rand Corporation, R.3324-ICJ (1985).

41. 1980 Report, p. 68.

42. J. Trauberman, "Statutory Reform of 'Toxic Torts': Relieving Legal, Scientific, and Economic Burdens on the Chemical Victim," *Harvard Environmental Law Review* 7, no. 180 (1983).

43. Ibid.

44. 1980 Report, table II–4, pp. 60–61.

45. I. Selikoff, *Disability Compensation for Asbestos-Associated Disease in the United States* (New York: Mount Sinai School of Medicine, 1983), p. xvi.

46. Barth and Hunt, *Workers' Compensation*, p. 181.

47. 1980 Report, p. 80; and Selikoff, *Disability Compensation*.

48. The reimbursement provisions create two kinds of disincentives for workers considering filing for compensation. The first disincentive is financial: a successful tort plaintiff must reimburse a workers' compensation insurer from his winnings. The second is procedural: a would-be tort plaintiff who has filed for workers' compensation must first secure the assent and cooperation of the relevant compensation insurers, even if they are controverting his claim. Thus, claimants and their attorneys are encouraged to pursue third-party recoveries whenever possible and to ignore workers' compensation. See, for example, Longshoremen's and Harborworkers Compensation Act, 33 U.S.C., Sec. 919 (1984).

49. Twenty-one states continue to impose restrictive definitions that limit coverage to diseases "peculiar to" or "characteristic of" a workers' occupation. Approximately thirty states continue to exclude "ordinary diseases of life," apparently surrendering to the etiologic problems raised by work-related diseases. See 1980 Report, pp. 68–69.

Arizona's statute typifies those that define occupational disease restrictedly. A claimant filing under the statute must meet a six-pronged "proximate causation" test in order to prevail:

1. There must be a direct causal connection between the conditions under which the work is performed and the occupational disease.
2. The disease must follow as a natural incident of the work as a result of the exposure occasioned by the nature of the employment.
3. The disease must be fairly traceable to the employment as its proximate cause.
4. The disease must not be the result of a hazard to which workmen would have been equally exposed outside of the employment.
5. The disease must be incidental to the character of the business and not independent of the relation of employer and employee.

6. The disease must appear to have had its origin in a risk connected with the employment, and to have flowed from that source as a natural consequence, although it need not have been foreseen or expected.

Ariz. Rev. Stat. Ann., sec. 23–901.01 (1983). *Accord* Mont. Code Ann., sec 39–72–408 (1983) (except for clause 6).

50. See, for example, Squillante's Case, 389 Mass. 396, 450 N.E.2d 599 (1983).
51. See Task Force, p. 48, and Barth and Hunt, *Workers Compensation*, pp. 267–68.
52. Barth and Hunt, *Workers' Compensation*, pp. 167–68.
53. For example: Ill. Ann. Stat., ch. 48, subsec. 1401–1420 (Smith-Hurd 1984).
54. For a discussion of the problems associated with employee monitoring with particular reference to the National Institute for Occupational Safety and Health (NIOSH) and Occupational Safety and Health Administration (OSHA) guidelines for monitoring toxics, see N. Ashford, C. Spadafor, and C. Coldart, "Human Monitoring: Scientific, Legal and Ethical Concerns," *Harvard Environmental Law Review* 8 (1984): 263.
55. See Berman *Death on the Job*, pp. 1–4.
56. Second injury funds are a standard recommendation to alleviate this threat to the employment of partially disabled workers. See generally Larson, *Workmen's Compensation*, 2, sec. 59.3(a).
57. Such an earnings record may be obtained by writing to the Social Security Administration, Bureau of Data Processing, Baltimore, MD, 21235, and by including the claimant's name, Social Security number, and a request for Type I information for the years desired.
58. Mr. Justice Holmes, while a member of the Supreme Judicial Court of Massachusetts, held that the opinion of an expert could be received, even if based on inadmissible sources, because the expert "gives the sanction of his general experience," *National Bank of Commerce v. City of New Bedford*, 175 Mass. 257, 261, 56 N.E. 288, 290 (1900), quoted in *Wing v. Commonwealth*, 359 Mass. 286, 290, 268 N.E.2d 658, 660 (1971). See also Trani's Case, 4 Mass. App. Ct. 857, 858, 357 N.E.2d 339, 340 (1976) ("A qualified expert may testify as to his opinion, even if the basis for that opinion is chiefly derived from inadmissible sources.").
59. Gots, "Medical/Scientific Decision-Making in Occupational Disease Compensation," in Task Force, appendix I.
60. See Task Force, pp. 68–75 (Recommendation J, Administrative Standards).

# 8
# Pollution Liability Insurance as a Mechanism for Managing Chemical Risks

## MARTIN T. KATZMAN

Public concern with the consequences of human exposure to toxic chemicals in the environment is at a historic high. Two recent federal statutes are explicitly concerned with exposures from hazardous waste disposal and chemical accidents: the Resource Conservation and Recovery Act (RCRA) and the Comprehensive Environmental Response, Compensation, and Liability Act (commonly called Superfund).[1] Both acts encourage the creation of a market in pollution liability insurance as a mechanism for regulating low-probability, high-consequences chemical risks.

The unique characteristics of chemical hazards make the creation of a market in pollution liability insurance extremely difficult. The uncertainties of scientific causation and the latency of response have given rise to the tort innovations discussed in Chapters 5 and 6. These legal changes pose a major challenge to traditional concepts of insurability.

This chapter examines the potential role of pollution liability insurance in victim compensation. The insurability of liabilities for toxic injuries is compared to conventional insurance exposures.[2] The evolution of the pollution liability market is examined in the context of the related "crises" in insurance and torts. Institutional reforms to stimulate the creation of a stable market are suggested.

### INSURANCE AS A RISK-MANAGEMENT REGIME

Two goals of societal risk management are deterring accidents and compensating victims.[3] American society has developed three approaches to managing risks: statutory regulation, common law, and market incentives. Statutory regulation, as exemplified by the Federal Insecticide, Fungicide, and Rodenticide Act, works through explicit constraints on behavior, in the form of prescriptions and proscriptions.[4] Activated *before* accidents occur, regulation focuses upon deterrence and places little weight on issues of compensation. Activated *after* an accident, the common law process attempts to discover the causal linkages when injuries result from torts or breach of contract. The common law apportions liabilities between injurer and victim for the primary purpose of just compensation; however, deterrence is often a guiding principle in the outcome.

Market incentives "internalize" into the economic calculus of a potential injurer the costs to third parties. In market processes, issues of blame or fault are irrelevant, but market mechanisms can be harnessed for purposes of both efficient deterrence and just compensation. Economists are virtually unanimous that user fees are the most efficient method for controlling third-party damages or "external diseconomies." While such a notion is widely rejected by the public as a "license to pollute," fees on pollution-engendering activities can encourage changes to less polluting processes and products.[5] Where third-party damages are random, like the catastrophic release of methyl isocyanate at Bhopal, then the appropriate user-fee mechanism is liability insurance.

Insurance has an interdependent relationship with the tort process. Liability insurance indemnifies the insured for legal liabilities resulting from injuries to second and third parties. Examples of such injuries are birth defects caused by a pharmaceutical (breach of contract) and illness caused by the contamination of a community water supply by leakage from a factory (a tort). The likelihood of an insured party being liable for such damages depends upon the tort rules.

The viability of the tort process depends in turn upon the availability of insurance. Even if a victim wins in court, recovery cannot always be assured. Where the damage is catastrophic from the viewpoint of the injurer, he may not have the means to compensate the victim. In several incidents, polluting businesses were dissolved before damages were discovered and the plaintiffs could sue. In other incidents, damages exceeded the net worth of the polluter.[6] Obviously, a business has no incentive to reduce the probability of accidents whose consequences become visible after dissolution or for which losses exceed net worth. Here the "financial responsibility" requirements come into play.

To reduce a defendant's ability to escape from indemnifying plaintiffs through bankruptcy or insolvency, legislatures have enacted financial responsibility requirements; a familiar example—automobile liability insurance is compulsory in most states. Congress has enacted financial responsibility requirements for risky business activities in the Clean Water Act, Trans-Alaska Pipeline Act, Deep Water Port Act, and Outer Continental Shelf Act, and as indicated in the Resource Conservation and Recovery Act.[7] The first four acts primarily concern the transportation of petroleum; the last two, the generation and disposal of hazardous waste. While not required, these acts encourage the purchase of insurance for both sudden accidents and "nonsudden occurrences."

While its primary purpose is risk spreading, the insurance market can potentially serve the deterrent function of risk management. If premiums are based upon expected damages, then insurance on risky processes or products becomes more expensive. This higher cost induces corporate risk managers to seek and adopt cost-effective measures for reducing risk. Furthermore, higher insurance costs for riskier products translates into higher product prices, which discourages demand. The ability of the product to compete in the market is an important test. If the product is consequently priced out of the market, then its risks are disproportionate to its benefits.

Traditionally, the insurance industry has played a major role in the management of conventional risks. Through the analysis of their own loss experience and experiments with risk-reducing technologies, the insurance industry has developed merit-rating practices for fire, boiler explosions, and industrial accidents.[8] By providing both the incentives for efficient deterrence and the means of just compensation, insurance may guide corporate risk managers in the direction of achieving more "acceptable risk."

## CONVENTIONAL INSURANCE AND CONVENTIONAL INJURIES

The conventional role of insurance is best understood in the context of commonplace, "mechanical" injuries and the conventional liability rules they entail. The commonplace injury is epitomized by the automobile accident. This accident is characterized by an individual victim, a defined injurer, clear causal links between the behavior of parties and the occurrence of the accident, and the immediate manifestation of the resultant injury.[9]

Compensation for the conventional injury can be sought through the tort or administrative system. Essential elements of the tort system, as it existed circa 1960, include the triggering of the statute of limitations at the time of occurrence of the accident, the negligence theory of liability, and an apportionment of liability among tort-feasors with respect to specific injurious acts.

Where the transaction cost of the tort system was perceived as excessive, it has been replaced by no-fault, administrative systems. As explained in Chapter 7, workers' compensation was adopted in virtually all states in the period 1910–30.[10] Similar administrative systems were set up for occupational disease, such as the Black Lung Fund for miners.[11] About fifteen states had adopted some form of no-fault rules for automobile accidents by the late 1960s.[12]

Whether adjudicated by a tort or an administrative process, mechanical injuries embody the characteristics of readily insurable risks.[13] First, they are homogeneous, numerous, and statistically independent enough to permit risk pooling. Second, losses are frequent enough so that a distribution of loss severities is calculable for purposes of premium setting. Third, the fact of loss is clearly determinable to avert false claims. Forth, the timing of a loss is easily ascertainable so that an insurer can close its books on a policy year and determine whether a profit or loss was made. Fifth, the magnitude of a single loss is rarely catastrophic from the perspective of the insurer's financial capacity.

### The Problem of Moral Hazard

Inherent in the risk-spreading function of insurance is the potential discouragement of efforts at taking due care and preventing accidents. The ability of the insured to transfer the financial costs of accidents to the insurer may decrease the incentive to minimize the probability of accidents ("loss prevention" in insurance jargon) or to mitigate the consequences of an accident once it occurs ("loss protection" in this jargon). Insurance may even encourage the insured to take steps to bring about an accident to collect insurance, like the owner of a dilapidated building who hires an arsonist. These perverse incentives create a "moral hazard," which is the most serious threat to the viability of conventional insurance markets.

Insurers attempt to mitigate the problem of moral hazard by several mechanisms. First, they may establish deductibles, which force the insured to pay for the "first dollar" of losses. Second, coinsurance requires the insured to bear some proportion of losses beyond the deductible. Third, there are policy limits, which reduce the incentive for bringing about a loss on an overvalued exposure. Finally, merit rating and risk classification strive to set premiums in proportion to the risk an exposure entails. It is through merit rating that insurance can serve public purposes in risk management.

### Conventional Lines

Historically, the insurance industry has based its growth on the expansion of "lines" of exposures that meet these criteria: marine navigation, human lives, fires, industrial accidents, illnesses, and automobiles. What about exposures that do not meet these criteria? Why cannot they be insured at some high price? Indeed, exceptional or unique exposures, like satellites and appendages of celebrities, are insurable by specialists like Lloyd's of London. In these cases, the losses are catastrophic from the viewpoint of the insured but not the insurer. The difficulty in analyzing the exceptional risk and the ambiguity of the loss distribution induces insurers to charge extremely high prices, thus thinning the potential market.[14]

The Comprehensive General Liability (CGL) policy has traditionally served as the first line of defense against liability for commercial ventures. Unless explicitly excluded, coverage is extremely broad. The insured presumes protection from all risks, including "unknown hazards," that may arise after the policy is written.[15]

## THE INSURABILITY OF CHEMICAL RISKS

The unique characteristics of chemical hazards make the creation of a market in pollution liability insurance extremely difficult. These include the persistence of many synthetics in the environment, the delay in manifestation of injuries from exposure, and the large number of potential causes of most diseases.[16]

These characteristics make proof of causation difficult and expensive. As indicated in Chapters 5 and 6, the rules of toxic torts have adapted to the changing nature of the risks. In particular there has been the displacement of the conventional statutes of limitation by the "discovery rule," the abandonment of negligence for strict liability, and the widespread adoption of new doctrines of joint and several liability. Both the inherent characteristics of chemical risks and the changing legal rules strain the conventional standards of insurability. Let us consider how the risks conform to these standards.

### Numerous, Homogeneous, and Uncorrelated

The number of exposed facilities is relatively large. In the United States, there are 115,000 chemical plants. About 67,000 are large enough to be included in the RCRA "manifest system" that records shipments of more than 100 kg. of waste per month. There are 5,000 transporters and at least 100,000 industrial waste disposal sites, of which at least 30,000 contain hazardous wastes.[17]

Producers range from large-scale producers of bulk chemicals to small specialty houses. While there are 65,000 chemicals in production, there are many similarities in engineering processes. The diffusion of RCRA standards of care should homogenize waste-handling practices.

Because chemical facilities are widely distributed around the nation, sudden accidents are likely to be uncorrelated. For example, a Bhopal-type explosion in California is unlikely to trigger a similar catastrophe in New Jersey. Nonsudden or gradual pollution may be another story. If a commonly used chemical like PCBs is discovered to be a carcinogen, all firms handling the chemical may become jointly and severally liable simultaneously. An insurer that covered all PCB producers faces correlated risks. Under RCRA, firms disposing of waste must demonstrate responsibility for $1 million per sudden accident and $3 million per nonsudden, or gradual, pollution incident. Not all chemical firms, however, are candidates for liability insurance. Under RCRA and Superfund, firms with considerable financial strength can self-insure. Specifically, firms possessing a net worth of more than $12 million can do without insurance. Responsibility can be demonstrated by several financial instruments, like letters of credit, but the most common form is pollution liability insurance. In fact, in 1985, the last year of expansion in coverage, nearly half of the firms with great financial strength purchased pollution insurance.[18]

A major source of correlation of losses is the commodity nature of most chemicals; that is, on a molecular level, they may be indistinguishable as to source of manufacture. If a given chemical X is discovered to cause latent environmental harm, all manufacturers might become liable under theories of joint and several liability. If an insurer concentrated its portfolio upon generators or handlers of that chemical, it would face the same correlated risk as an insurer whose portfolio consisted solely of hurricane damage on the Gulf Coast. An insurer could avoid this eventuality by insuring across many chemical products.

### Calculability of Frequency/Severity Distributions

Traditionally, the insurance industry has set premiums on an actuarial basis. To estimate the frequency/severity distribution, losses must be fairly common.[19] Contrary to public impression, actual losses from chemical injuries have been so rare and circumstances so unique that calculating the frequency and severity of losses from experience is virtually impossible. Even if historical loss data were available, they would reflect outmoded technologies, no longer in currency. This problem is characteristic of many complex, innovative technological systems, like satellites and offshore oil rigs. In the absence of actuarial data, insurers employ simple techniques of screening risks on the basis of such factors as proximity of facilities to water supplies.[20]

Instead, probabilistic risk-assessment methods, like event-tree and fault-tree analysis, can be used. Developed for weapons systems and nuclear power plants, these methods trace the chain of events that could unleash a catastrophe, like nuclear meltdown.[21] The chain of events that could result in a chemical catastrophe, however, is far more complex than anything that is conventionally insured. Far less is known about the toxicology of human exposure to chemicals than about exposure to radiation. Nevertheless, there have been promising attempts to apply risk-analysis techniques to chemical facilities.[22]

### Definiteness of Loss

Losses that cannot be publicly verified lend themselves to counterfeit claims. Property damage, ecosystem contamination, and many personal injuries, like tumors or birth defects, can be publicly validated. Until recently, public policy toward hazardous chemicals has focused exclusively upon such injuries.

As indicated by Root and Schnare in Chapter 4, the use of cancer as the epitome of environmental disease is problematic. Morbidity may take the form of a diffuse malaise, analogous to the debilitation associated with lead poisoning. The courts are increasingly responsive to claims for compensation for diffuse injuries. In *Ayers v. Jackson Township*, plaintiffs argued that a contaminated municipal water supply was responsible for malaise, rashes, and general anxiety.[23] While the trial court rejected such reports of malaise or rash as evidence of injury, it did award the plaintiffs $2 million for emotional stress and $8.2 million for lifetime medical monitoring. An appeal overturned the recovery for stress and medical monitoring, and the case now rests with the New Jersey Supreme Court. Regardless of this outcome, there is no guarantee that future courts will not recognize malaise as a compensable ailment in its own right. Indeed, a federal court of appeals held that employees may recover damage for both the reasonable probability of getting cancer and the resulting anxiety.[24]

### Determinability of Timing of Loss

Traditionally, the insurance industry has employed "occurrence" policies, which are activated at the time of an occurrence causing damage or injury. For mechanical technologies, where the temporal lag between occurrence and consequence was small, the claim was usually made in the year of the occurrence. At the end of the year, the insurer would know whether its premiums were sufficient to cover its losses. Premiums would be adjusted accordingly.

Several major oil spills in the 1960s raised insurers' awareness of a whole class of accidents that could have long-term environmental consequences. The time lag inherent in chemical technologies makes occurrence policies unacceptably risky from the insurer's perspective. Does the occurrence refer to the leakage of the chemical, the exposure of the plaintiff, or the manifestation of the injury? For a party covered by a sequence of insurers, it would be impossible to determine whose policy covered the occurrence. Furthermore, insurers would never know whether they could close the books on any of their policy years in the past. For the insurer, an occurrence-based pollution liability policy is basically unworkable.

These difficulties resulted in two innovations. First, Comprehensive General Liability policies written after 1973 excluded pollution occurrences, which were defined as "an accident, including continuous or repeated exposure to conditions, which result in bodily injury or property damage neither expected nor intended from the standpoint of the insured." This exclusion did not apply to sudden accidents, which were similar in temporal demarcation to mechanical accidents.[25] This exclusion created a gap in coverage. A second innovation, the Environmental Impairment Liability (EIL) policy was developed on the London market. As they evolved, EIL or pollution liability policies have been tailored to the unique characteristics of chemical hazards. Most important, they are claims-made rather than occurrence-based policies.[26] In other words, EIL policies are activated when the claim is made, which obviates the necessity of finding out when the accident occurred.

A claims-made policy is well suited to guaranteeing compensation to a victim. Because premiums are based upon expectation of past actions that could result in present claims, the claims-made policy is a poor deterrent. If a corporation abandoned the production of all suspected toxins and sanitized all of its current operations, it would not necessarily enjoy any corresponding reduction in premiums.

### Noncatastrophic Loss

The magnitude of potential losses should not be catastrophic from the perspective of the insurer's financial capacity. Insurers first confronted low-frequency/high-severity exposures in the nuclear power line. Congress hoped to establish a private nuclear liability insurance industry in order to provide compensation in the event of injuries. Insurers argued that damages could be so catastrophic as to jeopardize their solvency.[27] They feared that they could never collect enough in annual premiums to create an adequate reserve fund. While such fears have not been realized in the nuclear line during the past thirty years, similar fears are raised about pollution liabilities.

### Moral Hazard

The availability of insurance may reduce the incentive of the insured to take due care. Indeed, when the concept of pollution liability insurance was first suggested, both insurers and regulators thought it would create a serious problem in moral hazard. The belief was that pollution resulted from conscious business decisions and that the foreseeable consequences of conscious decisions were uninsurable.[28]

The moral hazard problem has been attenuated by preconditions on insurance and by regulation. To reduce the risks of moral hazard, the insurer may specify certain preventive measures as a condition for receiving insurance. These preconditions may be difficult for the insurer to enforce. Regulations are an alternative bulwark against moral hazard. Violations of regulations on technical performance or effluent limitations can void a liability policy. As a result, there is little talk now of insurance offering a license to pollute.

### Interrelationships among Injuries, Torts, and Insurance

The interrelationships among the characteristics of injuries, of the tort system, and of insurance contracts can be summarized at this point (Table 8.1). The classical mechanical injury and the emerging chemical injury are polar opposites on several dimensions (Panel A): the number of victims, the number of possible injurers, the preciseness of the loss, the clarity of causal links, the time lag between exposure to harm and manifestation of injury, and the frequency/independence of the accidents.

While the mechanical injury was amenable to compensation under the conventional tort rules, the chemical injury is not. Thus, the new toxic tort system differs from the conventional one on rules for triggering the statutes of limitations, burden of proof of causality, and the degree to which liability is collective or individual (Panel B).

The characteristics of both the injury and the tort system render the conventional insurance contract, the CGL policy, unworkable. The policies differ on

**Table 8.1**
**Strains on the Tort and Insurance Systems from Chemical Catastrophes**

A. Characteristics of Injuries

| Mechanical | Chemical |
|---|---|
| Individual victims | Mass victims |
| Defined injurer | Multiple injurers |
| Definite loss | "Fuzzy" loss |
| Clear causality | Multicollinearity |
| Sudden | Latent |
| Common/uncorrelated | Infrequent/correlated |

B. Characteristics of Torts

| Conventional torts | Toxic torts |
|---|---|
| Statute of limitation triggered by exposure | Statute of limitation triggered by discovery of injury |
| Negligence | Strict liability |
| Individual liability | Joint & several liability |

C. Characteristics of Insurance

| CGL | EIL |
|---|---|
| Exclude gradual pollution | Covers gradual pollution |
| Occurrence | Claims-made |
| Actuarial science | Risk analysis |
| Prospective/deterrent | Retrospective/poor deterrent |

three major dimensions. The event activating the policy (occurrence vs. claim) corresponds to the rules for triggering the statutes of limitations. Suddenly manifest injuries and short statutes of limitations are compatible with occurrence-based policies. Latent injuries and long "tails" on the statute of limitations are compatible with claims-made policies (Panel C). Thus, a dichotomy between general liability and pollution liability corresponds to the dichotomy between sudden injuries and gradually developing diseases.

Finally, the nature of the underwriting activity changes. Premium setting for the general liability policy, which insures common accidents, can be issued on the basis of actuarial analysis. For the pollution liability policy, which insures rare occurrences, premium setting must rely upon probabilistic risk analysis. Because the CGL policy covers the year of its issue, it is prospective in orientation and its premiums serve a deterrent function. Because the EIL policy covers past activity the consequences of which may surface in the year of its issue, it is retrospective in orientation and the deterrent function of its premiums are attenuated.

## THE RISE, FALL, AND RESURRECTION OF THE EIL MARKET

When RCRA was enacted in 1976, several London insurers had been developing liability policies for nonsudden or gradual pollution accidents. By the time Superfund was enacted in 1980, a few American insurers began offering such policies. Despite the misgivings of many underwriters, at least one dozen primary insurers were offering liability policies by 1983.[29] In addition, more than forty insurers had established a reinsurance pool to spread the risks further. It appeared that the federal initiative toward market-based regulation of chemical hazards might succeed.

By the end of 1984, the pollution-insurance initiative lay in shambles. London reinsurers had withdrawn from the market, carrying many existing and prospective insurers in their wake. Pollution insurers numbered about eight worldwide, most of whom insure small-scale facilities like gas stations and dry cleaning establishments. Only two insured the large-scale facilities and "heavy" risks, like chemical plants, from which catastrophes are most likely to result. By the middle of 1985, only one insurer offered coverage for nonsudden occurrences.[30] The most fortunate businesses have merely found their policy limits severely slashed and premiums raised. The unfortunate have had to scramble just to obtain premium quotations.

While there were some signs of recovery by the end of 1985, the pollution liability market has not developed as rapidly as the federal government had hoped. The financial responsibility requirements of RCRA and Superfund appear virtually unenforceable.

What has gone wrong? To what extent does the collapse of the pollution liability market reflect a specific problem with this line of insurance, a remediable imperfection, or merely a cyclical disequilibrium in general insurance markets?

### Conditions Affecting Market Viability

Unwritten criteria for insurability is the predictability of the tort process and the sanctity of insurance contracts. While the insurer can expect the tort law to evolve and contracts to be reinterpreted in unanticipated ways, judicial decisions in three areas have virtually undermined the predictability of insurance contracts. These are: (1) the adoption of theories of joint and several liability; (2) the interpretation of the pollution exclusion; and (3) the activation of liability policies for first-party damages.[31]

## Joinder of Defendants

As elaborated in Chapters 5 and 7, the courts have increasingly adopted theories of joint and several liability to apportion blame among defendants in cases of chemical victimization. The joinder of defendants (holding them liable as a group) increases the likelihood that victims of chemical injuries will be compensated by someone. While the joinder of defendants serves the goal of just compensation, it reduces the incentives for care. Indeed, a calculating polluter would benefit by being the sloppiest chemical handler in an industry since the savings would accrue to itself, but the costs of an accident would be shared by the industry. Undoubtedly, joint and several liability increases moral hazard.

The adoption of theories of joint and several liability raises the likelihood of an insurer's indemnifying clients for claims resulting from damages caused by other firms. In setting premiums, the insured can analyze the inherent frequency and severity of its own insured's losses. An insurer of chemical risks, however, cannot easily assess the risks of those firms with which its insureds might be joined in an action. Furthermore, the insurer can offer incentives to the insured for reducing risks. But the insurer has no contractual channel to influence the behavior of firms with which the insured may be joined. Clearly, joint and several liability reduces the insurability of pollution liabilities.

There are several partial solutions to the externalities resulting from the joinder of defendants. First, the development of regulations or self-imposed industry standards can reduce the insurer's risk for future damages from the joinder of a careful client to a careless competitor. The chemical industry generally favors the establishment of standards of care through regulation.[32] Insurers have an incentive to monitor the insured's adherence to both regulations and standards, because coverage lapses in the event of willful violation. Second, chemical manufacturers are voluntarily cleaning up abandoned sites.[33] This reduces future perils from past practices. Third, a waste-management industry has developed, which is specialized in transporting, storing, and disposing of waste. Many manufacturers are excavating chemicals buried on their own sites for relocation to these specialized sites, which will assume the first tier of liability.

Since joint and several liability can be viewed as mutual insurance de facto, the chemical industry might be amenable to more formal arrangements. The industry might establish its own mutual insurance pool, where its standards of care become preconditions for obtaining insurance. The rudiments of such a mutual market have been established by an asbestos removal firm and chemical waste handlers.[34] The mutual, retroactive assessments on nuclear power plants is a less relevant model to chemical risks. While nuclear power plants are relatively similar in size and technology and subject to common regulatory standards of care, chemical handlers are heterogeneous in these respects, and uniform retroactive assessments would be unfair and inefficient.

## Reinterpreting the Pollution Exclusion

The courts increasingly ignore the pollution exclusion by redefining gradual pollution as "sudden" and "accidental" from the standpoint of the insured's knowledge and intent.[35] In essence, the courts have transformed the CGL policy into a pollution liability policy with unlimited coverage. Because of the gradual nature of pollution, the limits of previous years can be activated *ad infinitum* once the coverage of one year has been exhausted. While such a layering of policies has not occurred so far, there is no guarantee that it will not in the future.[36] Nevertheless, the dilution of the exclusion has effectively depressed the market for EIL insurance, since many potential polluters may feel adequately covered by their CGL policy.

To obviate further confusion, the Insurance Services Office has tightened the pollution liability exclusion in the CGL policy. After 1986 there was no coverage for sudden and accidental pollution in this policy. On a prospective basis, this complete exclusion neatly partitions the market for pollution-related accidents from other risks. Contractual confusion about whether a particular incident is sudden or nonsudden is irrelevant in a consolidated, claims-made EIL policy. The new partition cannot overcome the unpredictability of CGL policies written before 1986.

## Indemnity for First-Party Cleanup

A liability insurance policy is intended to be activated by damage to third parties, not to the insured itself. EIL policies are not intended to indemnify insureds for on-site cleanup necessary to comply with the law. Nevertheless, the courts increasingly require insurers to indemnify polluters for cleaning up their own property on the grounds of preventing an imminent hazard to third parties.[37]

Insurers argue that in rewriting insurance contracts in an arbitrary way the courts are finding a "deep pocket" to finance a social program. The insurer's pocket is not as deep as the public believes. The $100–$200 billion estimated cost of cleanup is several times greater than the surplus of and annual premiums collected by the American insurance industry.[38]

The insurance industry argument is not totally convincing, however. It is true that few insurers anticipated either that they would be held liable for cleanup or for off-site pollution damages under old CGL policies. There may be cases where the cost of cleanup is cheaper than the expected costs of off-site damages. If on-site cleanup is not covered, the insured has an incentive to allow an off-site leakage to occur because it is insured, rather than to pay for an ounce of prevention. From the insurer's perspective, an on-site cleanup exclusion may prove

**Table 8.2**
**Innovations Induced by the New Toxic Torts**

| Phenomenon | Problem | Response |
|---|---|---|
| Joint & several liability | Moral hazard | Voluntary standards<br>Compulsory standards<br>Voluntary cleanup<br>Spinoff waste management |
|  | How to set rates? | Mutual insurance |
| Ignoring pollution exclusion | Financial losses | Insoluble? |
|  | Depress EIL market | Total exclusion in CGL |
| First-party cleanup | Financial losses | Consider in future rating |

to be cost-effective. Insurers might have to pay even more if the damage was permitted to occur.

In principle, there is no reason why all insurers could not knowingly underwrite policies on cleanup costs, as several do.[39] Insurers routinely offer policies on accidents that have already happened, by gambling on being able to charge a premium greater than the discounted settlement costs.[40] That insurers failed to collect sufficient premiums to cover their losses may reflect more on poor business judgment than poor public policy.

## Insurability and Induced Innovation

Innovations induced by threats to insurability caused by judicial behavior are summarized in Table 8.2. The diffusion of theories of joint and several liability creates problems of moral hazard and rate setting. The moral hazard problem can be addressed by voluntary or compulsory standards, voluntary cleanup, and spin-offs of waste-management activities. The rate-setting problem can be addressed by mutual insurance institutions, where achieving industry standards of care are prerequisites of insurability.

The courts' ignoring of the pollution-exclusion clause of the CGL policy has resulted in serious financial losses for insurers as well as depressing the market for EIL insurance. While the windfall losses to insurers are water over the dam, newly issued CGL policies contain a total pollution exclusion that should eliminate such losses in the future.

Requiring insurers to pay for first-party cleanup has also caused major financial losses to underwriters. In the future this eventuality can be factored into the premium, just as the costs of warranty repairs are factored into the prices of consumer products.

### Resuscitating the Market

The reduction in the number of pollution liability insurers is occurring at the same time federal and state financial responsibility requirements are being extended. As risk managers are demanding more insurance coverage, the supply is diminishing. How can this disequilibrium be resolved?

The collapse of the pollution liability insurance market in 1984 was partially a result of cyclical readjustments in reinsurance markets. The practice of setting premiums on the assumption of high-interest earnings proved disastrous as interest rates fell. Unexpectedly high losses resulting from large, mechanical accidents (like satellite losses or oil rig collapses) as well as large indemnity payments for toxic torts (primarily asbestos and pharmaceutical litigation) contributed to the loss of underwriting capacity.[41]

To some extent the shortage of reinsurance is self-correcting, as premiums rise and new capital flows into this sector. The current spurt in pollution liability premiums and reduction in availability of coverage may simply reflect this equilibrating process. Market volatility may also be an aspect of learning how to insure a new line, as suggested by the experience of the medical malpractice market in the 1970s.[42]

If the collapse of the pollution liability market was merely a cyclical or learning-curve effect, then the market will revive spontaneously. In this case, the proper public policy is to do nothing. So long as premiums are unregulated, the supply of insurance should be forthcoming. Indeed, several underwriting groups have recently returned to the market. From a nadir of eight firms at the end of 1984, the number of active insurers exceeded a dozen one year later.[43]

Problems inherent in analyzing environmental risks for purposes of rate setting are being resolved through the growth of specialized consulting firms. Problems of adverse selection and limited demand can be reduced by federal enactment of stricter financial responsibility requirements. For example, several states prohibit self-insurance and set the required insurance above the federal levels. These stricter financial responsibility requirements do not appear to have affected the attractiveness of these states from the viewpoint of either the insurer or the insured.[44]

There remain fundamental problems of insurability that result from legal risks, however. Toxic tort law is converging to a new equilibrium, but interstate variations in statutes and common law regarding joinder of plaintiffs are a major source of confusion. While legal variation has provided opportunities for experimentation, the marginal costs of further variation, after a decade or so, clearly surpass the marginal benefits. The most important step in resuscitating the market is increasing the predictability of insurance contracts and liability rules. Even an imperfect federal statute might be better than the current maelstrom.

If the pollution liability market fails to revive spontaneously, then states might create assigned risk pools. Insurers that wished to write CGL policies in a state may be required to write EIL policies in addition. A particular underwriter might refuse to do business in a state with such an assigned risk pool. For such pools to function, they would have to be established in virtually all states simultaneously.

Assigned risk pools would function in a regulatory capacity only if they were able to set premiums freely. If premiums were regulated on grounds of equity or affordability, then insurance would serve no deterrent function. Most state automobile liability pools provide poor examples of flexible premium setting, for they mandate cross-subsidies from good driver to bad. Concerns about chemicals, however, may eliminate pressures to make premiums "affordable."

Alternatively, the federal government might establish its own insurance program. Experience with such programs has not been encouraging. Federal deposit, crop, and flood insurance programs have suffered considerable pressures to equalize and subsidize the premiums.[45] A federal pollution insurance program would be subject to intense lobbying on the part of business to obtain similar premium subsidies. Since the principle "the polluter pays" is firmly entrenched in public policy, there may be little public support for such subsidies.

If none of these insurance strategies works, then the experiment in market-mediated regulation of risks will fail. The remaining approach is to unlink the problems of deterrence and victim compensation. Standards of care established by regulation would aim to achieve efficient deterrence. Either first-party medical and disability insurance or a quasi-public fund, like workers' compensation, would address the compensation issue.

First-party insurance for medical expenses and property damage is a workable risk-spreading device.[46] Insurers would not have to distinguish between environmental, occupational, or other causes. Injuries like pain and suffering, emotional distress, and birth defects are not insurable on a first-party basis. Requiring the victim to pay for her own insurance against pollution-engendered damages, however, would seem unfair or unjust. Publicly funded, catastrophic medical insurance might become a social program with uncontrollable costs.[47]

A congressionally mandated study suggested the creation of a two-tier compensation mechanism for toxic injuries. One tier consists of an administrative system for reimbursing medical payments and lost wages, based upon rebuttable presumptions. The compensation fund is to be financed by a tax on the production of chemicals, just like the clean-up activities of Superfund. The other tier consists of the new toxic tort law, with more formidable barriers to recovery balanced against the potential for large awards for pain, suffering, and other damages. Because of the lesser burden of proof, tier one would offer recovery to a larger number of victims than tier two, but at a considerably lower level of compensation.[48]

When Superfund was passed, the victim-compensation fund proposal was defeated. Opponents saw the attenuated burden of proof as offering an open-ended entitlement. Citing the experience of the Black Lung Fund for miners, they saw no grounds for excluding anyone with the remotest claim of injury from exposures to chemicals in the environment.[49]

Three states have established victim-compensation funds, and the fears of the opponents have yet to be realized. In California, for example, fewer than half a dozen claims have been filed.[50] As victims become aware of the existence of pollution victim-compensation funds, however, Pandora's box might open indeed.

## CONCLUSION

Pollution liability insurance was expected to play a major role in assuring victim compensation for chemical injuries. The passage of financial responsibility requirements, however, does not automatically create a private insurance market. While the growth of demand for such insurance has been rapid, inherent scientific uncertainties in risk assessment and the related vagaries of tort rules have undermined this market. The clamor for Congress to "do something" is not surprising. But will the proposals in Congress to support insurance institutions improve or worsen the situation?

## NOTES

1. Resource Conservation and Recovery Act, 42 USC 6924 (a) and (t) (1986); 40 CFR 264.14 and 265.14 et seq. Originally passed in 1976 and subsequently amended, this act deals with financial responsibility for treatment, storage, and disposal sites. Comprehensive Environmental Response, Compensation, and Liability Act, 42 USC 9608 (1986). Originally passed in 1980 and subsequently amended, this act aims to extend the financial responsibility requirements to generators of hazardous waste.

2. In insurance parlance, an "exposure" is an activity of a party that is subject to risk and hence to insurance.

3. Guido Calabresi, *The Cost of Accidents* (New Haven: Yale University Press, 1970).

4. Federal Insecticide, Fungicide, and Rodenticide Act, 7 USC 136 et seq.; 40 CFR 162–180.

5. Steven J. Kelman, "Economic Incentives and Environmental Policy: Politics, Ideology, and Philosophy," in *Incentives for Environmental Protection*, ed. Thomas C. Schelling (Cambridge: MIT Press, 1983), ch. 14.

6. Martin T. Katzman, *Chemical Catastrophes: Regulating Environmental Risk through Pollution Liability Insurance*, Huebner Foundation Insurance Series (Homewood, Ill: Richard D. Irwin, 1985), ch. 3.

7. Clean Water Act, 33 USC 1321 (p) (1986); Trans-Alaska Pipeline Act, 43 USC 1517(1) (1986); 33 CFR 132.1 et seq; Deep Water Port Act, 33 USC 1517(1) (1986); 33 CFR 137.301 et seq.; Outer Continental Shelf Act, 43 USC 1815 (1986); 33 CFR 132.1 et seq.

8. Herbert S. Denenberg et al., *Risk and Insurance* (Englewood Cliffs, N.J.: Pren-

tice-Hall, 1974), pp. 82–85; Robert I. Mehr and Emerson Cammack, *Principles of Insurance*, (Homewood, Ill.: Richard D. Irwin, 1980), pp. 346, 576; Mark R. Greene and James S. Trieschmann, *Risk and Insurance* (Cincinnati, Ohio: South-Western Publishing Co., 1984), pp. 199–201, 340–43.

9. William M. Landes and Richard A. Posner, "Tort Law as a Regulatory Regime for Catastrophic Personal Injuries." *Journal of Legal Studies*, 13 (1984):417-34.

10. Lawrence M. Friedman and Jack Ladinsky, "Social Change and the Law of Industrial Accidents," *Columbia Law Review* 67 (1967):50–82; James R. Chelius, "Liability for Industrial Accidents: A Comparison of Negligence and Strict Liability Systems," *Journal of Legal Studies* 5 (1976):293–310; and James L. Croyle, "Industrial Accident Liability Policy of the Early Twentieth Century," *Journal of Legal Studies* 7 (1978):279–97.

11. James D. Strader and Philip J. Sheehe, "Federal Black Lung: Ten Years of Legislation and Litigation," *The Forum*, 16 (1981):525–55.

12. Elizabeth M. Landes, "Insurance, Liability, and Accidents: A Theoretical and Empirical Investigation of the Effect of No-Fault Accidents," *Journal of Law and Economics* 24 (1982):49–66; Paul S. Kochanowsky and Madelyn V. Young, "Deterrent Aspects of No-Fault Automobile Insurance: Some Empirical Findings," *Journal of Risk and Insurance* 52 (1985):269–88.

13. Mehr and Cammack, *Principles*, ch. 2 and 4.

14. Robin M. Hogarth and Howard Kunreuther, "Ambiguity and Insurance Decisions," *American Economic Review* 75 (1985):386–90; Martin T. Katzman, "Creating Markets for Catastrophic Insurance" (Paper presented to the American Risk and Insurance Association, Chicago, Ill., August 1986).

15. Mehr and Cammack, *Principles*, ch. 13; Robert M. Tyler and Todd J. Wilcox, "Pollution Exclusion Clauses: Problems in Interpretation and Applications under the Comprehensive General Liability Policy," *Idaho Law Review* 17 (1981):497–521.

16. Talbot Page, "A Generic View of Toxic Chemicals and Similar Risks," *Ecology Law Quarterly* 7 (1978):207–44; Robert K. Best and James I. Collins, "Legal Issues in Pollution-Engendered Torts," *Cato Journal* 2 (1982):101–36.

17. Senate Committee on Environment and Public Works, *Injuries and Damages from Hazardous Wastes—Analysis and Improvement of Legal Remedies, Report of the Superfund Study Group* (97th Congress, 2nd sess.; 1982), ch. 2; Katzman, *Chemical Catastrophes*, p. 77.

18. Katzman, *Chemical Catastrophes*, p. 93.

19. For introductions to the estimation of loss distribution, see R.E. Beard, T. Pentikainen, and E. Pesonen, *Risk Theory: The Stochastic Basis of Insurance*, 3d ed. Monographs on Statistics and Applied Probability (London: Chapman and Hall, 1984); Robert V. Hogg and Stuart A. Klugman, *Loss Distributions* (New York: John Wiley & Sons, 1984).

20. Glenn R. Harris et al., "Groundwater Pollution from Industrial Waste Disposal," *Journal of Environmental Health* 44 (1985):386–90.

21. Norman J. McCormick, *Reliability and Risk Analysis: Methods and Nuclear Power Applications* (New York: Academic Press, 1981); Martin T. Katzman, "Environmental Impairment Insurance and the Regulation of Chemical Pollution," *CPCU Journal* 39 (1986):163–73.

22. E.M. Drake and E.S. Kalelkar, "Handle with Care: Using Risk Analysis for Hazardous Materials Facilities," *Risk Management* 28 (1981):44–50; Raymond F. Boy-

kin, Raymond A. Freeman, and Reuven R. Levary, "Risk Assessment in a Chemical Storage Facility," *Management Science* 30 (1984):512–17.

23. *Ayers v. Jackson Township*, N.J. Super. L., 189 N.J. Super 561, 461 A.2d 184 (1983).

24. *Jackson v. Johns-Manville Sales Corp.*, 54 U.S.L.W. 2400 (5th Cir. 1986).

25. Tyler and Wilcox, "Pollution Exclusion Clauses."

26. Matthew Lenz, Jr., *Environmental Pollution: Liability and Insurance* (New York: Insurance Information Institute, 1982); Katzman, *Chemical Catastrophes*, ch. 5.

27. Harold P. Green, "Nuclear Power: Risk, Liability and Indemnity," *Michigan Law Review* 71 (1973):479–510; Dan R. Anderson, "Limits on Liability: The Price-Anderson Act versus Other Laws," *Journal of Risk and Insurance* 45 (1978):651–74.

28. Lenz, *Environmental Pollution*.

29. U.S. Treasury, "Hazardous Substance Liability Insurance," March 1982; and William A. Mahoney, "A Risk Manager's Guide to Pollution Liability Policies," *Risk Management* 29 (1982):12–22.

30. "Difficulty in Obtaining Liability Insurance Said Major Problem for Waste Facilities," *Environmental Reporter* 15 (1985):1660–61.

31. Robert E. Keeton, *Venturing to Do Justice: Reforming Private Law* (Cambridge: Harvard University Press, 1969), p. 42; and Richard A. Schmalz, "Superfund and Tort Law Reforms—Are They Insurable?" *The Business Lawyer* 38 (1982):175–92; Bradford W. Rich, "Environmental Litigation and the Insurance Dilemma," *Risk Management* 32 (1985):34–43.

32. For summaries of congressional testimony by the Chemical Manufacturers Association (CMA), the trade association of large chemical companies, see "Limit Landfill Use, CMA Spokesman Urges," *Chemecology*, February 1983, p. 3, where more private recycling is favored; "Disposal Law May Need Changes, CMA Spokesman Tells Congress," ibid., May 1983, p. 12, which indicates that the chemical industry wants to fill gaps in RCRA and currently inspects common disposal sites to make sure that its standards are being followed; "Industry, Government Seek Solutions to Toxic Air Pollution Problems," ibid., July/Aug. 1983, p. 6, where the industry asks EPA to broaden the range of regulated emissions.

33. The initiatives of major chemical companies in voluntary cleanup is described in the organ of the Chemical Manufacturers Association; see "3M Funds Disposal Site Cleanup," *Chemecology*, September 1983, p. 10; "Chemical Company [Chevron] Voluntary Action Speeds Disposal Site Cleanup," ibid., November 1983, p. 5; "Monsanto Earmarks $25 Million for 1984 Waste Cleanup Program," ibid., April 1984, p. 4; "Industry Leader [Ciba-Geigy] Urges Voluntary Waste Cleanup," ibid., April 1984, p. 8. For a discussion of the role of large companies in the formation of Clean Sites, Inc., see "Speeding Hazardous Waste Site Cleanup—Industry, Conservationists Work Together," ibid., May 1984, pp. 2–3, "Environmental, Industry Groups Tackle Hazardous Waste Disposal Sites," ibid., July/August 1984, p. 2.

34. Stephen Tarnoff, "Asbestos Removal Firm to Form," *Business Insurance*, 3 June 1985; "House Panel, Association Announce Efforts to Resolve Environmental Insurance 'Circus'," *Environmental Reporter* 16 (1986):1766–67.

35. Kenneth A. Abraham, "Judge-Made Law and Judge-Made Insurance: Honoring the Reasonable Expectations of the Insured," *Virginia Law Review* 67 (1981):1151–99; Eugene R. Anderson and Avraham C. Moskowitz, "How Much Does the CGL Pollution

Exclusion Really Exclude?'' *Risk Management* 31 (1984):28–35; Rich, "Environmental Litigation."

36. Eugene R. Anderson, "The Proposed ISO General Liability Insurance Policy Revision," (Paper delivered at the Annual Meeting of the Risk and Insurance Management Society, New Orleans, La., April 1985).

37. Malcolm Aickin, "Environmental Impairment Liability Underwriting" (paper delivered at the Annual Meeting of the Risk and Insurance Management Society, New Orleans, La., April 1985).

38. In 1980 the surplus (the accumulated undistributed profits) of American property-casualty insurers was about $48 billion; the premiums written about $90 billion. See S.S. Huebner, Kenneth Black, Jr., and Robert S. Cline, *Property and Liability Insurance* (Englewood Cliffs, N.J.: Prentice-Hall 1982), pp. 514–15.

39. Donald V. Jernberg, "Insurance for Environmental and Toxic Risks: A Basic Analysis of the Gap Between Liability and Coverage," *Federation of Insurance Counsel Quarterly* 34 (1984):123–46; All-Industry Advisory Council (AIRAC), *Pollution Liability: The Evolution of a Difficult Market* (Oak Brook, Ill: AIRAC 1985).

40. Michael L. Smith and Robert C. Witt, "An Economic Analysis of Retroactive Liability Insurance," *Journal of Risk and Insurance* 52 (1985):379–401.

41. Leslie Cheek III, "Background Memorandum on the Availability of Environmental Liability Insurance and the Insurability of Environmental Liability" (mimeographed, Crum and Forster Insurance Companies, March 18, 1985).

42. Mehr and Cammack, *Principles*, pp. 80–82.

43. Robert A. Finlayson, "EIL Market Could Rebound," *Business Insurance*, 29 April 1985.

44. States where insurers find adverse legislation generally are not those with compulsory insurance requirements or higher minimum requirements. See AIRAC, "Pollution Liability," table 5.

45. Mehr and Cammack, *Principles*, p. 267; Fred B. Power and E. Warren Shows, "A Status Report on the National Floor Insurance Program," *Journal of Risk and Insurance* 46 (1979):61–76.

46. See the argument of Patricia M. Danzon, "Tort Reform and the Role of Government in Private Insurance Markets," *Journal of Legal Studies* 13 (1984):517–49.

47. Richard Zeckhauser, "Coverage for Catastrophic Illness," *Public Policy* 21 (1973): 149–72.

48. Senate Committee, "Injuries and Damages"; Stephen M. Soble, "A Proposal for the Administrative Compensation of Victims of Toxic Substances Pollution: A Model Act," *Harvard Journal of Legislation* 14 (1977): 683–824.

49. See Strader and Sheehe, "Federal Black Lung."

50. Robert S. Faron, "EIL Financial Responsibility Comes of Age," *Risk Management* 30 (1983):20–32.

# 9
# The Common Law under Challenge: Politics, Science, and Toxic Torts

## JAMES M. STROCK

One of the key issues that has confronted Congress in recent years has been how, and in what situations, to respond to calls for legislation by various interests seeking federal intervention in the resolution of claims for compensation for injuries alleged to result from exposure to toxic substances. In the 99th Congress, there was particular focus on toxic-compensation issues during reauthorization of the Comprehensive Environmental Response, Compensation, and Liability Act (CERCLA, also known as "Superfund") and the Price-Anderson Act, as well as proposals concerning the liability of manufacturers of vaccines and product liability generally.[1]

This chapter will examine these and other legislative responses to the "compensation conundrum," and will view such "solutions" in the context of the political realities in which Congress operates. To accomplish this, we first discuss the manner in which toxic compensation issues arise in Congress.

## THE POLITICAL CONTEXT

### Victim Assistance as a Political Matter

Elected representatives in Washington tend to be responsive to two general types of constituent appeals. The first includes situations in which a constituent group, which may be politically organized, seeks specific legislative aid to meet a defined objective. A second type of appeal is one that may be less specific in its immediate objective, and may have less advocacy from organized interest groups, but nonetheless represents a major current in public opinion that must be considered by elected officials. Issues of toxic compensation have garnered both types of constituent appeals. Initially, organized appeals from discrete groups focused attention on parts of the issue—seeking federal legislation providing compensation for injuries from exposure to nuclear radiation from federal weapons testing, asbestos, improperly disposed hazardous waste, or the defoliant Agent Orange, which was used by the American government in the Vietnam conflict.

At the same time, a general public concern has arisen over the issue of toxic hazards in the environment. Elected officials have encountered such concerns resulting from specific problems in their home states and localities, and polling data has confirmed the emergence of this issue. For example, reliable data collected in 1984 indicated that, in one state, respondents listed environmental issues as their top concern—ahead of issues such as employment, which usually is stated to be most important in all regions of the nation.[2] While such a development could, to some degree, be explained by a greater awareness of a broad range of environmental issues, there is no doubt that public awareness of issues relating to toxic substances was a key factor in this and other polls. And there is equally no doubt that political figures in all regions of the nation are receiving similarly urgent signals.[3]

## An Emerging Consensus in Favor of Federal Action

As public attention has become focused, a consensus has begun to emerge that federal action is required. As a result, members of Congress from all regions and all parts of the political spectrum are examining potential legislative responses. In addition to compensation proposals from liberal members who are predisposed to national solutions, one also finds conservatives prepared to embrace federal solutions. For example, Senator Hatch of Utah, a leading proponent of private sector primacy and state sovereignty, has sponsored a leading proposal to compensate one alleged class of victims; another conservative, Senator Kasten of Wisconsin, has been the primary sponsor of a measure that would preempt state tort law in the related area of product liability.[4]

At the same time as the consensus in favor of federal action has grown in strength, various factors have made it likely that eventual legislation will deal with toxic compensation issues in a piecemeal manner, rather than comprehensively. One apparent reason is the nature of the "consensus" itself: although many congressmen agree that a national response is required, they by no means agree on the goals or content of such a response. Some would use federal authority to limit the liability of enterprises alleged to be the sources of injurious toxic exposures, while others would expand potential liability beyond that now provided in state tort law.

Another factor militating against a comprehensive response to toxic compensation is the institutional context in which such issues arise in Congress. Because the implications of targeted compensation programs could be far-reaching if applied across-the-board, most proposals have dealt with what appears to be a limited, well-defined part of the problem. Senator Hatch, for example, the sponsor of legislation to ease legal recovery for injuries alleged to result from nuclear weapons testing in Utah, cannot be assumed to similarly support generic toxic compensation programs.[5] On the other side, Senator Mitchell, a liberal who is likely less concerned about the implications of a generic approach, has also sponsored piecemeal, targeted reforms as the best available alternative among a limited array of options.[6]

In turn, because the issue has been fragmented, it enters congressional consideration in shards. At various times, compensation proposals have been considered in the Senate committees on Environment and Public Works, Veterans' Affairs, and Judiciary. Other committees would be likely to request referral of any comprehensive approach, including Finance, and Labor and Human Resources. When one takes into account the even more complex jurisdictional situation in the House of Representatives and the numerous executive branch departments and agencies that would also be impacted, it becomes clear that enactment of a comprehensive approach would face significant obstacles. Indeed, without focused presidential leadership, such an approach must be considered highly unlikely at least in the absence of a budgetary, legal, or environmental crisis of major proportions.[7]

## CONGRESS—THINKING LIKE LAWYERS

No one should be surprised that Congress views the toxic compensation debate from a legal perspective; by far, most of its members are lawyers.[8] However, there is a deeper and more significant reason why the legal perspective arises as the dominant viewpoint.

The tort system has been, throughout our history, the primary means through which compensation for injured individuals has been provided, as well as the main mechanism for affixing liability for such injuries. As several commentators have ably argued, the legislative branch (at the state level as well as the federal) has interacted with the judicial branch on such issues in the face of the perceived inability of the common law of torts to balance competing interests consonant with new realities of commercial interaction and public perception.[9] So it was during the industrialization of America, and so it is today, in a time when scientific knowledge of the relative risk of various activities is dramatically increasing, simultaneous with the development of new legal tools designed to expand the reach of the tort system.[10]

As a result, toxic compensation issues have entered the political debate under the more general rubric of mass torts, and it is worthwhile to briefly review the various ways in which toxic torts have come to the attention of Congress.[11]

## Traditional Mass Torts

As previously described by Billauer and Locke (Chapters 5 and 7), the tort system will work effectively if the proper evidentiary basis is set. Without reentering the debate into the always elusive substantive versus procedural distinction, it is nonetheless accurate to suggest that traditional mass tort actions can resolve claims arising from the immediate aftermath of discrete disasters, because the substantive elements of tort law are either those of system delivery or the determination of the extent of compensation that an enterprise should be required to provide. Such questions are difficult and may be contentious, but they are the traditional questions of tort law, and if causality is not at issue, compensation is likely to arise. Indeed, it is the familiarity of such questions that has encouraged some observers to suggest that toxic compensation issues can best be resolved by further procedural reforms in the context of mass torts.[12]

## A Third Toxic Tort

Two types of toxic torts have been discussed at length in Chapters 5, 6, and 7. These are the time-lag torts, where the onset of harm arises well after the initial exposure, and the multiple-causation torts, where a class of individuals may be affected but compensation for a single individual may be limited due to other potential causes of the eventual harm. There is a third situation that has been sketched out by Katzman in Chapter 8, and that reflects more an institutional perspective than a matter of the tort system. These are torts seeking damages in situations of absolute uncertainty of both causation and exposure source. In the past, one would expect no standing to sue without a reasonable causation argument. The success of toxic torts, however, has opened up this new tort—if a disease that can be associated with a toxic exposure arises, then sue any and all potential defendants. The alleged injuries from exposure to hazardous waste sites are examples of this tort. Generally, such torts share the following characteristics:

- uncertainty as to the identity or solvency of potential defendants;
- uncertainty as to the potential synergistic effects of various hazardous substances working together;
- the possibility that a release may have triggered an immune system disorder, causing the victim to fall prey to diseases normally unrelated to exposure to the specific substances in question;
- inability of the legal system to arrive at a reasoned judgment as to which of several hazardous substances can be considered to be the "cause" of injury to one individual;
- uncertainty as to the link between a specific defendant and the substance alleged to have harmed a plaintiff;
- uncertainty as to the duration and pathways of exposure in individual cases;
- lack of cooperation from defendants and potential third-party defendants fearful of their liability under the Superfund strict, joint and several liability standard;[13] and

- occasional ambiguity as to the extent a group alleging injury can be said to have "accepted the risk of injury."[14]

At the time of passage of the Superfund statute in 1980, in the aftermath of the highly publicized Love Canal incident in New York, Congress became embroiled in controversy as to how—and whether—to compensate individuals alleging injury from uncontrolled hazardous waste sites. In order to ensure that the new statute would be passed by the lame-duck Congress following the 1980 election, and to avoid the possibility that incoming President Reagan might veto the Superfund legislation upon taking office, a compromise was struck on the issue of victim compensation. That compromise, embodied in Section 301(e) of the statute, called for a study "to determine the adequacy of existing common law and statutory remedies in providing legal redress for harm to man and the environment caused by the release of hazardous substances into the environment."

The resulting study reflected several tendencies that have marked the debate on toxic compensation issues in Congress generally.[15] First, the entire study was predicated on the assumption that tort law, with specified reforms, could provide an adequate and fair mechanism for compensation. This was an outgrowth of the preceding debate, which focused on the advisability of new, federal tort remedies, and the fact that the Section 301(e) Study Group, as it came to be known, was composed entirely of lawyers.[16]

As a result, the 301(e) Study Group defined its mission in a manner that, to a great extent, predetermined both the results of its deliberations and the future terms of the debate. The group considered the adequacy of existing legal remedies for putative victims of hazardous substance exposure, but it did not attempt to quantify the need for any additional regime for such compensation or the relative effect such exposure, when considered as a whole, might be having on public health generally. In skating over such questions, the *301(e) Report* made its tort law reform recommendations in a vacuum of sorts, and did not include an extended examination of the overall regulatory and compensatory framework, within which any new regime for aid to a limited group of toxic victims would, of necessity, operate.[17] While recognizing that traditional tort law rules might not be adequate in this new area, it simultaneously assumed that the Superfund toxic compensation issue should be viewed, as previous, more traditional compensation issues had been, as a mass tort question.[18]

## CONGRESSIONAL COMPENSATION PROPOSALS

Viewing toxic compensation issues through the prism of toxic tort analysis, and spurred by general public concern and organized constituent pressure arising from specific incidents, members of Congress have proposed numerous federal fixes. These proposals can be placed in several categories.

### Federal Limitation of Enterprise Liability

There is ample precedent in American history for limitation on legal liability for new and emerging industries, going back to the Industrial Revolution in the nineteenth century.[19] Thus it is not surprising that proposals to limit the toxic tort liability of various enterprises have been prominent in recent years.

*Price-Anderson Act.* In order to explicitly provide support to the nuclear power industry, Congress provided, in the Price-Anderson Act, for the partial preemption of state tort law.[20] Specifically, in the event of an "extraordinary nuclear occurrence," state and common law defenses of federally licensed nuclear facilities are waived, while the overall liability of such facilities in the event of a mass disaster would be limited to a defined amount.[21]

While this limitation of liability may be viewed as especially important because it deals with one of the most visible areas in which toxic tort liability could arise, it is important to note several factors. An "extraordinary nuclear occurrence" connotes an accident more analogous to a traditional mass disaster, in which the act and the consequence are closely linked in time, and causation, as a matter of fact and law, is direct. The type of exposure and injury found in toxic torts—often including a time delay between exposure and ultimate injury—could arise in the context of subthreshold nuclear accidents as well as "extraordinary nuclear occurrences." In light of a recent Supreme Court decision holding that the Price-Anderson Act does not preempt state tort law in such subthreshold accidents, the preemption, insofar as it would deal with toxic torts, is incomplete.[22]

During the 99th Congress, there were several attempts to remove the general liability limits of the act, but none was successful. It is not unreasonable to conclude that the present limited liability regime for federally licensed nuclear facilities will remain in its current bifurcated state, or will be eroded further rather than expanded.

*Federal Reform of Product Liability Law.* Spurred by the spread of what they regard to be proplaintiff statutory and common law developments in many states, various business interests have made the federalization of product liability law a high priority on their Washington agenda.[23] In the 99th Congress, Senator Kasten led the effort to pass sweeping changes in this area of tort law, but was ultimately unsuccessful in winning Senate approval.[24]

Key concerns animating this reform effort have included the expanding liability of manufacturers and users of toxic substances and the availability of compensation for workers and others alleging injury as a result of exposure to such substances. In this area, as in others, there will be major confrontation between those interests seeking to continue to use the tort system as an evolving means for providing compensation for victims and placing costs on risk producers, those seeking an alternative to the tort system, such as an administrative system, and those seeking to limit tort liability while not providing an alternative compensation mechanism.[25] Inevitably, the implications of the product liability debate are leading business interests to pay increased attention to related areas in which toxic torts arise, such as environmental law.

*Indirect Liability Limitation through Generic Compensation.* To the extent that Americans suffer illnesses for which they have been placed at significantly increased risk as a result of toxic exposure, it is likely that they are receiving, or may in the future receive, compensation from a number of generic health care mechanisms. These mechanisms would include private medical insurance (personal and employer-provided health and disability coverage) and social insurance (such as Medicare for older people, Medicaid for the indigent, federal catastrophic health insurance, and state and local programs). As a result, increased generic coverage, particularly for older people, can be assumed to cover toxic-related illnesses that may have been caused or made more likely by risk producers who would, at least theoretically, be liable under tort law. By providing for compensation in such circumstances, the generic compensation mechanisms are, in a sense, subsidizing risk-producing activity in the absence of a funding mechanism that takes risk production into account.

An example of such indirect liability limitation may arise in the context of proposals for federal catastrophic health insurance, as was introduced in the 100th Congress by Senator Kennedy. However, in the current budgetary climate, large-scale expansion of social health care mechanisms is not likely, as is indicated by the absence of cradle-to-grave national health care as a major political issue in recent years.

## Federal Expansion of Enterprise Liability beyond State Tort Law

The 99th Congress saw various proposals to directly increase the toxic tort liability of business enterprises beyond that provided in state common law and statutes. Not surprisingly, such efforts focused on the reauthorization of the Superfund statute.

Although the CERCLA Section 301(e) Study Group considered expanding potential toxic tort liability in the Superfund context through creation of a federal cause-of-action in tort for individuals alleging injury resulting from exposure to toxic materials from uncontrolled hazardous waste sites, their eventual report to Congress did not endorse the concept. Instead, the report proposed a series of reforms to state tort law designed to remove procedural barriers to recovery. However, by pointing out such potential barriers, the group did lay a partial predicate for those urging federal efforts to expand liability beyond the state law.

Subsequent proposals for a federal cause-of-action have varied in their particulars and have been offered for application in several environmental statutes.[26] There are, however, several recurring elements of special importance, each of which relates to the problem of establishing the causal linkage between a release of a toxic substance and an injury to an individual.

Although some have urged that the burden of proof as to causation be shifted to the defendant after a threshold showing of exposure by the plaintiff, as in the *Allen* case, a recent proposal considered on the floor of the House did not include such a frontal assault on the causation question.[27] The primary sponsor of the legislation, Congressman Frank, acknowledging that "some provisions in . . . [other proposals'] language might result in excessive liability for those with minimal or no actual responsibility for injuries," proposed that plaintiffs establish, by a "preponderance of the evidence" that "the release caused the incurrence of the damages." In turn, the amendment would have made major changes in the Federal Rules of Evidence, relating to the admissibility and relevance of medical information offered to meet the required showing of causation: "Any information *which tends to establish* that exposure to a hazardous substance in question causes or contributes, or does not cause and contribute, to damages compensable under this title *of the type or class allegedly suffered by an individual* shall be treated as relevant evidence in an action under this title."[28]

The bill then listed specific types of information that would be considered relevant under its definition, notably including epidemiological studies and others designed to demonstrate increased risk among defined population groups rather than in individual cases.

The Frank proposal also imposed a federal "discovery" statute of limitations, in line with that of many states, while additionally providing for recovery of specified damages prior to enactment of the bill. Finally, it encouraged the use of the class-action mechanism of the Federal Rules of Civil Procedure, even though its use in the mass tort context is, at best, a matter of great controversy.

In the end, the Frank amendment was defeated.[29] However, the political compromises said to have provided the margin of defeat, as well as the implications of the adjustments to state statutes of limitations that were accepted in its stead, indicate that the federal cause-of-action is by no means a dead issue.

### Government-Operated Administrative Mechanisms

Given the historical familiarity of administrative compensation mechanisms, it is natural that Congress would consider them in the context of toxic exposures. Prominent examples include the state-run workers' compensation system, the controversial federal Black Lung Benefits Program, the Federal Employees Compensation Act, and the Veterans' Compensation and Pension Program.[30] Additionally, attention has been given to the Japanese Pollution Fund and a little-used California administrative system created for individuals alleging injury from exposure to hazardous waste.[31]

An administrative mechanism might expand or limit the liabilities of defendant enterprises, depending on its scope (would it serve as an alternative to tort remedies?), its funding sources (would it be paid for by the private sector alone, and which parts of the private sector?), and the rules by which decisions on compensation would be made in individual cases. As a result, partisans of all sides in the debate have looked to administrative solutions in the context of specific types of toxic-related injuries.

*CERCLA Section 301(e) Report Proposal.* As a supplement to the state tort law system, the Section 301(e) Study Group recommended the establishment of a state-administered administrative compensation system as part of the Superfund. The state programs, to be funded by a federal levy similar to the existing Superfund tax, would provide specified damages for individuals not otherwise compensated. Decisions on individual cases would be made on the basis of proof of exposure and subsequent development of a disease linked to such exposure by "Toxic Substance Documents."

Although various legislative proposals to incorporate this sort of administrative mechanism into CERCLA have been offered, none has become law. Criticism from industrial interests tends to focus on the use of presumptions. It is argued, with the simpler case of the Black Lung Program as precedent, that any such presumptions would be malleable, so that overcompensation would likely result as political pressure from organized groups of alleged victims is mobilized.[32] Further, the use of presumptions to avoid legal determination of causation in individual cases where such a determination is virtually impossible, rather than routine, has been well questioned by some lawyers and is discussed at some length in Chapter 5. While this conceptual difficulty with the use of presumptions exists generally, it is most relevant in a "type three" toxic tort, where causation may prove to be an almost insurmountable obstacle to recovery in court.

On the other hand, some groups representing individuals alleging illness from toxic exposures have been very hesitant to entrust their compensation to a branch of the same federal government that they believe has failed them in allowing exposure in the first instance. Others, while willing to believe that an administrative compensation system might provide meaningful assistance, have urged that damages be commensurate with those available to successful plaintiffs in court. At a time when the damages recoverable by a successful plaintiff are themselves a matter of great controversy, such suggestions have served to polarize debate.[33]

*Veterans' Disability Compensation—Agent Orange and Atomic Radiation.* In recent years, there have been a high number of claims from veterans seeking compensation for injuries alleged to have resulted from nuclear weapons testing between 1945 and 1963 and the use of the defoliant Agent Orange during the Vietnam conflict. In both categories, the Veterans Administration (VA) generally rejected disability claims, citing the inability of claimants to demonstrate that the injuries in question were "service connected," or that the link between exposure and the onset of long-latency disease or injury was tenuous as a matter of science. In the face of mounting political pressure, and in the midst of some uncertainty as to the potential federal government liability in such cases, Congress passed the Veterans' Dioxin and Radiation Exposure Compensation Standards Act, which was signed by President Reagan on October 24, 1984.[34]

In 1981 Congress had altered existing VA regulations limiting the provision of VA medical care to veterans claiming injury from exposure to Agent Orange during their service in Vietnam. However, disability compensation was usually not provided before the passage of the 1984 legislation. The legislation directed the VA to develop regulations for the resolution of both Agent Orange and nuclear-related claims for compensation, and required that such regulations consider whether presumptions of service-connected injury should be developed for specific diseases. Additionally, the VA was directed to give due regard to existing policy providing the benefit of the doubt to individuals' claims for service-connected diseases and disabilities for which "there is an approximate balance of positive and negative evidence"; to evaluate scientific information related to the link between exposure and injuries and disease, and to provide temporary compensation in certain cases.[35] The new law did not provide for judicial review of the decisions made in individual cases.

Contemporaneous litigation, in which Vietnam veterans filed a class-action suit against seven chemical companies that manufactured Agent Orange under government contract, was settled out of court on May 7, 1984. Under the terms of the settlement agreement, the defendant companies agreed to pay $180 million into a fund for veterans and dependents of veterans who suffered damages from exposure to the defoliant. In a candid statement highlighting the controversy surrounding this product-liability litigation, the presiding federal district court judge wrote, "There are serious factual problems with plaintiffs' case, the chief one being doubt that present scientific knowledge would support a finding of causality.[36]

*The Atmospheric Nuclear Testing Compensation Act.* In response to the allegations of individuals in several states, Senator Hatch of Utah introduced legislation to provide compensation, through an administrative mechanism, for injuries and disease linked to open-air nuclear testing in Nevada during 1951–62 and to uranium mining in Colorado, New Mexico, Arizona, and Utah during 1947–61.[37]

Compensation would be provided through a Commission on Compensation for Atmospheric Nuclear Testing that would be established to evaluate individual claims, and directed to:

set forth what [it] will define as the sizable possibility [that the cancer of an individual resulted from exposure to ionizing radiation] necessary to allow for an award of damages to individuals covered by this title; . . . determine how [it] will prorate the amount of actual damages other than damage for individuals with a probability of causation of less than 50 percent.[38]

In allowing compensation to be granted in cases unable to meet the 50 percent causation requirement of the legal system, the bill implicitly recognizes the limitations of the *Allen* decision, which dealt with the same nuclear-testing toxic tort scenario. Seeing that the U.S. District Court's rationale, if upheld on appeal, could nonetheless not be relied upon to provide a predictable means of compensation because of the causation question, Senator Hatch attempted to provide more certainty of recovery at a lower level of probability of causation, but within some apparently reasonable limits:

> In assessing claims for radiogenic cancer, the Commission shall rely upon the preponderance of current scientific opinion regarding the causal link between ionizing radiation and cancer, and the relationship between a given prior dose of radiation and the likelihood that a given cancer resulted from such dose. As guides to this scientific opinion, the Commission shall use only the relevant publications of authoritative scientific agencies and societies [a list follows].... The Commission shall then take into account the uncertainties that are entailed in determining whether a given dose of radiation caused a given cancer when those uncertainties are described in such publications. The Commission shall give claimants the benefits of the doubt regarding such uncertainties.[39]

In addition to that generic "benefit of the doubt," the bill also provided for the commission to grant the "benefit of the doubt" to individuals resident "immediately downwind" from the Nevada nuclear test site, or who were actually at the test site or working in specified uranium mines during the defined period.

In arguing for his compensation system, to be paid from a $150 million fund from the federal treasury, Senator Hatch admitted that "even if a definitive study could be performed and it did show that the downwind population had radiogenic cancers, these cancers would be hidden among many naturally occurring cancers. The study could not distinguish which cancer victims were the radiation victims."[40] Despite this difficulty, the Senator supported compensation for the downwind population, estimated to number from 172,000 to 200,000, on the basis that the failure of the U.S. government to monitor the radiation doses that individuals received from fallout had made understanding of the health effects unnecessarily unclear.

Opposition to the legislation, which was successful in defeating it in the Senate during the first session of the 99th Congress, focused on the "benefit of the doubt" that would accrue to an individual claimant solely on the basis of his or her membership within a larger group that may have incurred an increased risk of disease.[41] There was likely additional concern, not fully reflected in the floor debate, about the financial implications for the federal treasury: the amount provided in the amendment appears very low when compared against the number and types of claimants likely to be encountered, the relatively broad definition of compensable damages included, and the fact that the individual cases could be reopened during the fifteen-year term of the proposed law.

*Administrative Compensation in General.* Administrative, no-fault mechanisms have also been suggested, by various affected interests, for compensation for injuries from vaccines, asbestos, and manufactured products.[42] However, in the case of toxic torts, they provide no easy answers, because they cannot escape the question of causation in individual cases. The use of presumptions can regularize the processing of claims, but guarantees over- or undercompensation generally, and unavoidably allows for the application of political pressure. This is inevitable, because the scientific evidence may be more relevant to the question of group health effects than individuals' health effects—so the calibration of the presumptions is, inevitably, adjusted to meet nonscientific criteria. Further, the adoption of administrative mechanisms ensures that public and political attention will focus on those enterprises creating the risk in question, which may not result in fairness among other, similarly situated groups needing compensation, or an efficient allocation of scarce societal resources.

### Federal Insurance Proposals

In light of the political and practical difficulties attendant upon attempts to institute a traditional-type administrative system or a federal cause-of-action in tort, some in Congress have urged adoption of an insurance mechanism as an alternative. In theory, an insurance mechanism offers the advantage of aiding individuals as part of a larger group alleging exposure, and provides incentives for focusing on preventive health care and screening in addition to medical care after a health effect is manifest.

*The Mitchell "Victim Assistance" Demonstration Program.* Senator Mitchell proposed, as part of the effort to reauthorize the Superfund in 1985, a pilot program that would make available defined insurance benefits to individuals who, as part of a larger population group determined by health effects studies, face "significantly increased risk [of disease or illness] as a result of a release of a hazardous substance."[43] The proposal was based in part on a concept developed by the Keystone Center, and reflected an intensive effort by Senators Mitchell and Simpson to develop a limited program that would better define the "victim problem" while not, at the same time, setting into motion an "entitlement program that would be politically impossible to limit if found unnecessary or unnecessarily expensive."[44] Although, in the end, Senator Simpson felt compelled to oppose the Mitchell proposal, which was defeated in a close Senate floor vote after earlier approval by the Environment and Public Works Committee, the legislation nonetheless represented an important new current in the toxic-compensation debate.[45]

Section 129 of S.51 would have established a series of grants, not to exceed a total of $30 million per year, for a cumulative total of $150 million over five years, from the general revenue portion of the Superfund, to be provided to states seeking to establish and operate pilot compensation programs. States could be granted funding, at the sole discretion of the administrator of the Environmental Protection Agency, for "any geographic area... for which a health assessment or other health study performed under section 104(i) indicated" that

(a) there is a disease or injury for which the population of such area is placed at significantly increased risk as a result of a release of a hazardous substance; (b) such disease or injury has been demonstrated by peer-reviewed studies to be associated (using sound scientific and medical criteria) with exposure to a hazardous substance; and (c) the geographic area contains individuals within the population who have been exposed to a hazardous substance in a release.[46]

After the selection of an area for inclusion, the state would commence and administer a program of specified medical assistance for individuals within the affected populations, for a period of not less than three years or longer than five years. Such programs would be required to offer "appropriate medical screening, examination and testing" to determine whether individuals manifested the disease or injury for which the larger population was at "significantly increased risk." Then, for individuals "with no present symptoms of such disease or injury," participation in a group medical benefits policy would be offered, to pay for continued health monitoring. Those "with present symptoms" would be eligible for additional benefits, including reimbursement for previously incurred "out-of-pocket medical expenses in connection with such disease or injury," and a "group medical benefits insurance policy providing the reasonable costs of sound medical and surgical treatment and hospitalization."[47]

Opposition to the proposal took two forms: criticism relating to the perceived possibility that the program would grow out of control, despite the intended limits, and questions relating to the practical difficulties of operating such a program on a piecemeal, experimental basis. Those making the first point viewed the history of the Black Lung Program as illustrative of a general tendency of compensation programs to gather unassailable political strength. Reminded of the early, excessively low cost estimates offered by proponents of that program, such opponents suggested that the relatively restrictive compensation envisioned here would not prove to be a sufficient barrier against such a tendency.[48] They also were concerned that the studies by which the geographic areas would be defined were not themselves sufficiently delineated. Underlying each of these points was an overall sense that any sort of compensation system dealing with "type three" toxic exposures could not, by definition, separate elevated risk in individual cases from that of a larger population group.

Opponents also questioned the practical difficulties inherent in the arbitrary distinctions that a limited pilot program would present. For example, the definition of the geographic area within a single state could foreseeably lead to situations in which citizens of one state would receive major health benefits designed to deal with "significantly increased risk" of catastrophic illness, while others, equally exposed to a release of a hazardous substance—even the same release—would receive no such aid solely because they lived in an adjoining state. And even if a geographic area were to be defined in a generally accepted manner, would there by any distinction drawn between individuals who were in the area for shorter or longer periods of time? The ultimate question of societal fairness was also invoked: on what basis, given the scientific uncertainty surrounding causation, could Congress reasonably choose to provide compensation for a discrete group of victims when others, similarly situated, might receive no such aid?

*General Catastrophic Insurance.* Mindful of the societal implications of a limited compensation effort, but attracted to the other advantages an insurance mechanism might theoretically bring, some observers have suggested consideration of broader social insurance, which would include toxic injuries. Such an approach might differ from traditional social insurance programs if it was financed by imposing costs on those enterprises believed to be passing on involuntary, unanticipated risks on individuals.

One such alternative, which could supplement rather than supplant the tort system, might be a national program of secondary catastrophic medical insurance—similar to that envisioned in the Mitchell bill and modeled on existing state secondary insurance programs.[49] Such a program could be financed by those risk-creating enterprises that would otherwise face greater tort liability. One way to design such a system would be to centralize the federal government's quantitative risk-assessment efforts (related to more than thirty statutes) to produce data on comparative risk, for the purpose of assessing fees on risk producers.

Although no proposal along these lines is before Congress at the time of this writing, the confluence of several legislative trends indicates that it may become viable in the immediate future. In addition to the growing concern about the potential effectiveness of the legal system in the area of toxic torts generally (largely expressed by industrial interests facing staggering liabilities and transaction costs), there is a growing awareness of the interrelationship between toxic risks incurred in the workplace and those in other settings, including communities geographically proximate to hazardous waste sites or to enterprises that routinely handle toxic substances.[50] To the extent that the increased attention leads to greater knowledge of the nature of long-latency disease, it is perhaps fair to assume that public debate will begin to shift toward an explicit recognition of the relative risks connected with various activities and enterprises. The first measure of this is the first comprehensive analysis of relative risks performed by EPA, which documents an inverse relationship between actual risk and the public's perception of what is risky.[51]

Indicative of this tendency is the introduction of federal legislative proposals that would build upon state "worker right-to-know" and "community right-to-know" statutes. Under recent amendments to the Occupational Safety and Health Act, workers must be notified of exposures to hazardous substances that increase their risk of contracting various illnesses, and such workers must then be monitored, as part of a larger exposed group, for health effects.[52] The Superfund Amendments and Reauthorization Act dramatically increases the health effects data that must be developed by handlers of toxic substances and shared with geographically proximate communities.[53]

## LOOKING TO THE FUTURE

All can agree that any attempt to predict the future actions of the federal government on major policy issues is fraught with difficulty and must be laced with caveats. At the same time, the development of the toxic-compensation issue thus far can lead to reasonable suppositions concerning the future.

Initially, it is important to emphasize that toxic-compensation issues will likely continue to be viewed primarily as *legal* questions. Such issues are also likely to remain fragmented and be considered in relative isolation in various congressional committees, each focusing on parts of the larger societal problem, even though the clear interconnections between worker and community toxic-compensation issues, for example, are increasing. At the same time, the defeat of major compensation proposals during the Superfund debate in the first session of the 99th Congress revealed widespread and bipartisan congressional reluctance to create new administrative programs, based on experience in other areas, as well as a degree of contention among potentially affected interests that was so great so as to have led to temporary stalemate.

The advent of the era of the Gramm-Rudman-Hollings Deficit Reduction Act has also added a major new factor to the equation: it will be, for the foreseeable future, virtually impossible to institute new domestic programs, no matter how worthy, during a time when existing programs are being cut back dramatically.[54] Indeed, the 1988 House Appropriations Committee based its EPA funding proposal on the philosophy that there should be no new programs until there is a showing of greater risk avoidance through the new programs than by existing programs. The House also indicated that future EPA budgets should include a statement on the level of risk that will be avoided through each program element, in juxtaposition with the budget request.[55]

The overarching political reality of tight budgets enhances the attractiveness of providing compensation by means of new legal remedies for parties alleging injury—remedies that can easily be tailored to specific types of exposure. Of course, any such nonbudgetary fix might impose significant transaction costs on private parties. But some observers, glossing over the causation problems attendant in toxic compensation questions, are likely to argue that a federal cause-of-action would ensure that "the polluter pays."

Meanwhile, other developments may increase the attractiveness of this option. In particular, the development and dissemination of health effects data under various statutory authorities may provide information useful to plaintiffs in tort litigation, and may speed consideration of procedural changes such as those envisioned in the 1985 Frank amendment.

While such an expansion of legal remedies would trouble some, perhaps they can take limited comfort in conceding that this approach, while politically active, will continue to face the same institutional obstacles blocking generic application that face other proposals. And, if the results of such an approach are as negative as some suggest, it might trigger an economic or legal crisis of sufficient dimension to cause the executive branch to frame the toxic-compensation issue in the comprehensive manner that may be most appropriate.

## NOTES

The views expressed herein are solely those of the author, and should not be read as representing those of the Office of Personal Management, the U.S. Environmental Protection Agency, the Committee on Environment and Public Works, or any specific member of the committee.

1. 42 U.S.C., subsec. 9601–9659 (1982); P.L. 85–256, 71 Stat. 576 [codified as amended at 42 U.S.C., subsecs. 2012, 2014, 2039, 2073, 2210, 2232, 2239 (1982)]; and see, for example, S. 827, 99th Cong. 1st sess., 1985, and S. 100, 99th Cong. 1st sess., 1985.

2. See Edward J. Rollins, Memorandum to James A. Baker III concerning President Reagan's visit to New Jersey (October 1984) (unpublished material). While the high level of anxiety about toxic materials and the environment in New Jersey can be attributed in part to the many hazardous waste sites identified in that state, public awareness generally has increased as a result of television news coverage and fictional accounts of toxic exposures. In the aftermath of the Bhopal tragedy in India, for example, the ABC television network broadcast of March 2, 1986, a two-hour presentation entitled "Acceptable Risks," depicting a lifelike situation in which an American chemical plant, located in the United States, released toxins into the environment, killing hundreds of people in nearby neighborhoods.

3. A recent Harris Poll found the following environmental attitudes among the American people: 93 percent of adults believed that "pollution of lakes and rivers by toxic substances from factories" is a serious problem; 69 percent favor passage of a $10 billion extension of Superfund; and 16 percent said that they would not vote for a congressional candidate who took environmental positions differing from their own. *Inside EPA*, 4 July 1986, p. 12.

4. Amendment to S.J. Res. 77, *Cong. Rec.*, 99th Cong., 1st sess., 1985, 131:S15582–84. See also, S. 100, 99th Cong., 1st sess., 1985, sec. 3.

5. Senator Hatch did, however, oppose the successful effort of Senator Roth to delete the victim assistance provisions of the Senate Superfund reauthorization bill in 1985. See *Cong. Rec.*, 99th Cong., 1st sess., 1985, 131:S12004.

6. S. 51, *Cong. Rec.*, 99th Cong., 1st sess., 1985, 131:S11564–77.

7. This trend, while unfortunate, appears to be all but inevitable. Charles Harr has called this mode of policy development the "catastrophe theory of planning." Richard B. Steward and James E. Krier, *Environmental Law and Policy* (Indianapolis: Bobbs-Merrill, 1978), p. 332, quoting C. Harr, *Land-Use Planning* (1950), pp. 130–131.

8. According to the Library of Congress, Congressional Research Service, *99th Congress Profile* 1 (1985), "As has been true in previous Congresses, law is the dominant profession. . . . There are presently 251 Members who have listed law as their profession." In contrast, there is only one member who has a Ph.D. in basic science, and he is a minority member of the House, and is not on a health-related subcommittee (Ritter-PA).

9. See generally Morton J. Horowitz, *The Transformation of American Law, 1780–1860* (Cambridge, Mass: Harvard University Press, 1977).

10. See P. Huber, "The Bhopalization of U.S. Tort Law," *Issues in Science and Technology*, Fall 1985, p. 73.

11. This categorization of toxic torts was previously presented in J. Strock, "Coming to Terms with the Compensation Conundrum," *American Bar Association Journal*, August 1985, p. 68.

12. See, for example, D. Rosenberg, "The Causal Connection in Mass Exposure Cases: A 'Public Law' Vision of the Tort System," *Harvard Law Review* 97 (1984):851.

13. 42 U.S.C., sec. 9607. Although liability under the act is not strict, joint, and several on its face, numerous judicial decisions have interpreted the statute to provide for such liability. See, for example, *U.S. v. A & F Materials Co., Inc.*, 578 F. Supp. 1249 (S.D. Ill. 1984); *U.S. v. Wade*, 577 F. Supp. 1326 (B.D.Pa. 1983); *U.S. v. Northeastern Pharmaceutical & Chemical Co.*, 579 F. Supp. 823 (W.D.Mo. 1984).

14. This is not to suggest that assumption of risk would necessarily be available as a defense to a strict liability claim. See W. Prosser and W. Keeton *The Laws of Torts*, 5th ed. (1984), sec. 68 (defining assumption of risk). However, as a practical matter, the difficulty of segregating causation in individual cases (between various carcinogenic agents, taking into account voluntary and involuntary risks assumed, synergistic effects, etc.) makes it foreseeable that the law may, as a practical matter, consider such issues in toxic tort cases.

15. Senate, Committee on Environment and Public Works, *Injuries and Damages from Hazardous Wastes—Analysis and Improvement of Legal Remedies: A Report to Congress in Compliance with Section 301(e) of the Comprehensive Environmental Response, Compensation, and Liability Act of 1980 (P.L. 96–510) by the Superfund Section 301(e) Study Group*, 97th Cong., 2d sess., 1982, S. Rept. 12 [hereinafter cited as *301(e) Report*].

16. The advisability of a federal cause-of-action in tort was one of the key questions dividing members of the Senate Environment and Public Works Committee during the Senate deliberations that led to the creation of the Superfund. Such a provision was included in the Stafford bill, S. 1480, 96th Cong., 2d sess., 1980, section 4(a). The study of the issues provided under Section 301(e) was part of a compromise struck when the federal cause-of-action was deleted.

17. A draft study prepared for the use of the Environmental Protection Agency, *Estimated Uncompensated Costs of Selected Illnesses* at ES-1 (1984) attempted to begin such an analysis by gathering data concerning "the availability of compensation for health care costs and lost wages for people suffering illnesses that might be related to hazardous substance exposure. . . . Information on whether there is a need for additional compensation and, if so, the magnitude of that need would help to inform the debate on several recent legislative proposals in the Congress that aim to improve the access of victims to compensation."

18. The *301(e) Report* did mention existing nonlitigation compensation mechanisms already in use, such as the workers' compensation system. See *301(e) Report*, pp. 219–20. However, the report did not directly confront the question of whether a legal approach (with requirements of causation, etc.) would necessarily be preferable to, say, a social insurance alternative. Neither did it directly confront the complexities that would be faced in distinguishing among causes of disease in individuals who might have received workplace exposures or similar exposures other than those arising from hazardous waste sites. However, the report implicitly acknowledged such difficulties by attempting to craft procedures that would take into account the effects of new mechanisms for recovery of damages in the context of hazardous waste exposure.

19. See, for example, Limited Liability Act of 1851, c. 43, 9 Stat, 635 [codified as amended at 46 U.S.C., sec. 181 (1958)] (limiting liability of vessel owners to encourage shipbuilding).

20. Price-Anderson Act, P.L. 85–256, 71 Stat, 576 (codified as amended at 42 U.S.C., subsec. 2012, 2014, 2039, 2073, 2210, 2232, 2239 [1982]).

21. 42 U.S.C. 2210 (1982).

22. See *Silkwood v. Kerr-McGee Corp.*, 464 U.S. 254–256 (1983).

23. S. 100, 99th Cong., 1st sess., 1985, sec. 3(b)(1), provided for preemption of state product liability laws: "This Act supersedes any State law regarding recovery for any loss or damage caused by a product to the extent that this Act establishes a rule of law applicable to any civil action brought against a manufacturer or product seller for loss or damage caused by a product, including any action which before the effective date of this Act would have been based on (A) strict or absolute liability in tort; (B) negligence or gross negligence; (C) breach of express or implied warranty; (D) failure to discharge a duty to warn or instruct; or (E) any other theory that is the basis for an award for damages for loss or damage caused by a product."

24. See S. 100, 99th Cong., 1st sess., 1985, which provided an elective system of administrative compensation for product liability cases.

25. For historical precedents of this last approach, *see supra* note 19.

26. Amendment to H.R. 2817, *Cong. Rec.*, 99th Cong., 1st sess., 1985, 131:H11574–75.

27. Congressman Frank discussed his amendment described in the month before its introduction, referring to a "current draft" of the bill offered on December 10. See Amendment to H.R. 2817, *Cong. Rec.*, 99th Cong., 1st sess., 1985, 131:E5183.

28. Amendment to H.R. 2817, *Cong. Rec.* 131:H11574 (emphasis added).

29. The Frank amendment was defeated on a roll-call vote of 261–162. *Cong. Rec.* 131: H11585–86.

30. As of 1983, all fifty states administered workers' compensation programs. See U.S. Chamber of Commerce, *Analysis of State Workers' Compensation Laws 1983* (1983), p. 1; The federal compensation program for victims of pneumoconiosis began in Title IV, Federal Coal Mine Health and Safety Act of 1969, P.L. 91–173, 83 Stat. 792–98 (codified as amended at 30 U.S.C., subsec. 921–32 [1986]; Federal Employees Compensation Act, 5 U.S.C., subsec. 7902, 8101–8193 (1982); See, for example, Veteran's Disability, Compensation, Housing, and Memorial Benefits Amendments of 1981, P.L. 97–66, 95 Stat. 1026 (codified at 38 U.S.C., subsec. 101, note, 210, 314, 362, 411, 413, 414, 717, 752, 767, 769, 777, 802, 906, 1003, 1712, 1776, 1803, 1819, 1826, 1828, 1902, 1903, 3010, 3103A, 3203, 5010 [1982 and supp. 1986]).

31. For descriptions of this law, see B. Aronson, "Review Essay: Environmental Law in Japan," *Harvard Environmental Law Review* 7, no. 135 (1983): 150–71; and California Health and Safety Code, subsec. 25370–82 (West 1984 and supp. 1986).

32. The use of formal rule-making procedures, with rights of judicial review, might calm some such critics, but it should be remembered that, in addition to regulatory adjustment of presumptions, there is also the alternative of subsequent statutory adjustment, as was performed by Congress in the context of compensation for black lung disease in 1972, 1978, and 1981. For an overview of that program and the tortuous path of its legislative development, see J. Lopatto, "The Federal Black Lung Program: A 1983 Primer," *West Virginia Law Review* 85 (1983):677.

33. In particular, there has been great controversy surrounding the award of punitive

damages and efforts have been made to limit their application in the product liability context. See supra note 23, S. 100, Section 12.

34. See, for example, *In re Agent Orange*, 597 F. Supp., p. 879 app.; and, P.L. 98–542, 98 Stat. 2727 (codified at 38 U.S.C., sec. 354 [supp. 1986]) [hereinafter Veterans' Act].

35. As part of the new law, Congress found that "there is some evidence that chloracne, porphyria cutanea tarda, and soft tissue sarcoma are associated with exposure to certain levels of dioxin found in some herbicides and that most types of leukemia, malignancies of the thyroid, female breast, lung, bone, liver, and skin, and polycythemia vera are associated with exposure to certain levels of ionizing radiation." Veterans' Act, sec. 2(5). See also sec. 5(b)(2)(A)(i); sec. 2(13), 6, and 9.

36. *In re Agent Orange*, 597 F. Supp., p. 775.

37. Amendment to S.J. Res. 77, *Cong. Rec.*, 99th Cong., 1st sess., 1985, 131:S15582–84.

38. Ibid., p. S15583.

39. Ibid., p. S15584.

40. Ibid., p. S15586 (statement of Senator Hatch).

41. The Hatch amendment was tabled and thereby defeated by a roll-call vote of 53–42. Ibid., p. S15599.

42. See S. 827, 99th Cong., 1st sess., 1985; S. 1265, 99th Cong., 1st sess., 1985; and S. 100, 99 Cong., 1st sess., 1985, sec. 3.

43. S. 51, *Cong. Rec.*, 99th Cong., 1st sess., 1985, 131:S11570.

44. Keystone Center, *Potential Approaches for Toxic Exposure Compensation: A Report on the Conclusions of a Keystone Center Policy Dialogue* (1985).

45. Senator Roth's amendment to delete the Mitchell victim-assistance provision was accepted by the full Senate on a roll-call vote of 49–45. See Cong. Rec., 99th Cong., 1st sess., 1985, 131:S12004.

46. S. 51, *Cong. Rec.* 131:S11570.

47. Ibid., p. S11571. The group medical benefits insurance policy was subject to "an annual deductible of $500, with no copayment requirement or annual or lifetime limitation on expenditures other than those referred to in paragraph (3)." The deductible amount does not appear to have been carefully examined to gain an understanding of its likely consequences.

48. See *Cong. Rec.* 131:S11929 (statement of Senator Simpson).

49. See, for example, Ind. Code Ann., subsec. 27–8–10–1—27–8–10–8 (Burns 1986) ("Comprehensive Health Insurance"); Wis. Stat. Ann., subsec. 619.01–619.18 (West 1980 & Supp. 1985) ("Risk Sharing Plans"). See also Blue Cross & Blue Shield, Model Health Insurance Pooling Mechanism Act (1984) (model state legislation for the establishment of health insurance pooling mechanisms for "uninsurables").

50. For a summary of "worker right-to-know" and "community right-to-know" proposals at the state and federal levels, see B. Burby, D. Lawson and S. Woolum, *Right-to-Know: An Issue Analysis* (Government Research Corporation monograph, 1986).

51. See U.S. Environmental Protection Agency, *Unfinished Business* (1987). This is a five-volume report on the environmental risks that have not yet been addressed through regulation, and the relative importance of these risks.

52. Occupational Safety and Health Act, P.L. 91–596 (1970), 29 C.F.R. 1910.

53. Superfund Amendments and Reauthorization Act (SARA), Title III (Emergency Planning and Community Right to Know Act of 1986), 49 U.S.C. 11001 et seq.

54. Balanced Budget and Emergency Deficit Control Act of 1985, P.L. 99–177, 99 Stat. 1037 (1985).

55. H.Rept. 100–189, *A Report of the HUD and Independence Agencies Subcommittee of the Appropriations Committee to Accompany the 1988 HUD Appropriations Bill*, which includes the EPA appropriation.

# 10
# Conclusion

## MARTIN T. KATZMAN AND DAVID W. SCHNARE

Our story began with the fictional case of Mary Peters. Fortunately, we can report that while you were perusing this volume, she gave birth to a healthy baby girl. But the Peters family is not home free. They still face the risk that exposures to hazardous chemicals will result in injuries that are only manifested in the future. As a society, we have to decide whether the compensation options available to families like the Peters are adequate to meeting the goals of societal risk management.

## OPTIONS FACING THE INDIVIDUAL

Let us now synthesize what we have learned about the questions posed in the introduction. The questions have been reduced to three:

1. What harm results from exposure to toxic chemicals?
2. What medical treatment is available?
3. How should she seek compensation?

### What Harm Results?

The concept of the disease hierarchy introduced in Chapter 2 suggests that the existence of harm is not determinable on a yes or no basis. There are degrees of harm, ranging from a body burden (like the PBB buildup in the Michigan population), to malaise (affecting the population in Jackson Township, New

Jersey), to disease sentinels, to familiar symptoms of morbidity and mortality. Some of these harms are latent rather than manifest.

The question of what is the harm is central to the problem facing the attorney, the underwriter, and the regulator. To the attorney, the existence of a harm is prior to the question of determining its cause. Traditionally, tort law evolved to deal with easily verifiable harms, such as severed limbs, disability, or mortality. Only such overt conditions were potentially compensable, not speculative "fear of future harm." Improved knowledge of the disease process, however, has enhanced the medical community's awareness of sentinels of latent diseases and of more diffuse diseases such as reduced neurological function. The increased willingness of courts to define conditions lower down the disease hierarchy as harms expands the range of compensable conditions.

Billauer et al. applaud the unwillingness of courts to award damages for fears that are not associated with physical injury. To compensate for fear of possible disease is to engage in prospective risk assessment in the genre of the regulator. While it may be legitimate for social policy to adopt worst-case assumptions about risk in pursuit of the deterrence objective, Billauer et al. question the viability of this approach in pursuing the compensation objective. Indeed, the tort process traditionally demands a much higher standard of evidence than that a particular cause might remotely cause a harm. To allow compensation for chemical injuries on remote possibilities is to challenge the basis of civil law which weighs the preponderance of the evidence.

Medical monitoring for disease sentinels may be a method of reconciling the plaintiff's demand for compensation for fear and the inadmissibility of intangible mental evidence. The presence of verifiable disease sentinels might become essential for any plaintiff having standing to sue. Even if body burdens of chemicals can be detected, they may not inevitably result in harm. The plaintiff must still engage in subliminal toxicology, but the ground rules are clearer. Instead of attempting to prove a causal link between exposure and harm, the plaintiff must prove a link between body burden and harm.

To the underwriter, the changing definition of harm, more than its ever-expanding limits, makes the pricing of insurance extremely difficult. Whether the definition of harm adheres to the narrow, traditional constraints of manifest, definable injury or whether it encompasses more intangible harms like neurological impairment is not consequential to the underwriter. If the existence of neurological damage is verifiable and its extent measurable, then the magnitude of expected losses is calculable. What causes difficulty is the unpredictability of the boundaries of harm. If harm includes fear of harm itself, then the liability for harm is virtually unbounded.

The regulator has focused his efforts on protecting the public against carcinogens. The large-scale exposure of the public to chemicals that may not be carcinogens but that may cause harms lower down the disease hierarchy suggests that a refocusing of effort may be in order.

# CONCLUSION

Similar harms can result from different chemical causes. Not all chemical injuries are marked by such clear-cut biological fingerprints as berylliosis or mesothelioma. Overlapping symptomatology makes drawing of direct causal links between specific chemical exposures and biological consequences difficult, especially for most cancers.

Establishing causal links between chemicals and harm is not necessary in providing compensation for special damages. Indemnification for medical expenses, lost income, and loss of life can be provided by first-party insurance. Compensation simply requires the establishment of the fact of harm. General damages for pain and suffering, loss of consortium, or disfigurement, however, are not insurable on a first-party basis. For both compensation for general damages and for efficient deterrence, establishing causality is essential.

Regulatory approach focuses entirely upon deterrence, to the exclusion of compensatory objectives. As suggested by Cothern and Schnare, regulatory judgments about potential harm from hazardous chemicals are soft. Regulatory findings about chemical hazards are at best indirect inferences from very long chains of reasoning and extrapolations far beyond the limits of existing data. Each link in this chain is tenuous.

Regulatory incentives are biased toward avoiding the worst-case harm. The standards of evidence for risk of harm are quite different from "preponderance of evidence" used in tort actions. In setting standards, the regulator prefers to err on the cautious side. Consequently, the public tends to confuse remote possibility with great likelihood.

Can we expect regulators to protect us from harm? Not necessarily, because their attention is focused on new risks and new chemicals, to the exclusion of older risks that have been grandfathered out of scrutiny. The new risks may replace more dangerous old risks, but this issue of risk substitution is generally out of the regulator's purview. An analysis of risk substitution involves engineering assumptions about process substitution and behavioral judgments about product substitution.

Common law approach to establishing causality has both deterrence and compensation objectives. Unlike the regulator who has to make judgments about risk prior to the occurrence of harm, the tort system enjoys the benefit of hindsight. Nevertheless, there are formidable difficulties in extricating one cause from another. For example, is the plaintiff's neurological impairment a result of chemical exposure or of aging? Is nervousness due to chemicals or to stress?

Like the regulator, the attorney has no alternative to probabilistic reasoning. While the regulator is concerned with making judgments for populations, attorneys traditionally make arguments about individuals on a case-by-case basis. Billauer et al. question whether the same conservative assumptions used in setting regulations are appropriate to post facto compensation. This practice will inevitably lead to mistakes of overcompensation. It is not clear, however, what alternatives exist for plaintiffs.

## What Treatment Is Available?

The answer to this medical question has profound implications for the tort system, regulators, and insurance. Vague symptoms resulting from chemical exposure may lead a patient to seek out medical treatment. As Root and Schnare note, a large fraction of the population might experience such symptoms from the buildup of halogenated hydrocarbons. Fortunately, the subclinical buildup of fat-soluble residues in adipose tissue may be reversible. The two-step treatment (mobilization-excretion) now costs about $3,000, and costs may fall further.

The availability of treatment places a burden on the victim and establishes a limit on compensation. To compensate individuals exposed to chemicals in the environment, the courts have increasingly provided for medical monitoring. Such monitoring can allay the fear of disease if the substance is excretable and hence the condition reversible.

Unfortunately, such simple treatments are not applicable to substances like asbestos or coal dust, which are not stored in adipose tissue. There is no practical means of interrupting the latent process of carcinogenesis. Treatment is painful, and not highly successful for lung diseases.

Regulators have not considered the treatability of chemical harm, but Schnare and Root imply that perhaps they should. To the extent that conditions are reversible or treatable at a low cost, regulators should take this into account in risk analysis.

Underwriters have an interest in sponsoring research on the applicability of detoxification treatments. The availability of treatment provides a cap on potential awards for fears of chemical injuries. Instead of viewing such exposures as open-ended, actuaries can calculate the cost of medical monitoring and treatment in the case that body burdens are discovered.

## How Should She Seek Compensation?

Do all harms merit compensation? Traditionally, there has been a presumption in the common law to "let losses lay where they fall." The victim would cope with potential losses by "saving for a rainy day." This strategy is feasible for minor injuries such as auto repairs, but not for catastrophic injuries of the type considered here. In the latter case, an individual could never accumulate sufficient funds to cover the magnitude of the loss. As noted by Strock, both tort law and statute have evolved away from this traditional presumption to one of risk spreading.

The three systems available to Mary Peters for risk spreading are: (1) first-party insurance, (2) no-fault, administrative compensation, and (3) tort compensation. From the victims' viewpoint, these three alternatives differ in important attributes, such as timeliness and certainty of compensation, degree to which victim recovers for all damages, and transactions costs. None of the three compensation systems clearly dominates on all attributes.

# CONCLUSION

First-party insurance is generally available for property damage, medical expenses, loss of income from disability, and death. It is activated without fault by the fact of an injury. First-party insurance has the merits of rapidity and certainty of compensation and moderately low transactions costs. It covers special or economic damages, but not general damages like "pain and suffering," which are insurable.

A second form of risk spreading is compensation on a no-fault basis from the party able to control or spread the risk. Examples are the generic workers' compensation system and its cognates, like the black lung compensation system. If a victim of chemical injury suspects that exposure occurred in the occupational arena, then this is an appropriate avenue of recovery. Under workers' compensation, general damages are excluded.

The workers' compensation system is hardly automatic, since over 60 percent of the disease claims were litigated and only 15 percent of the plaintiffs won their litigation. Aside from black lung disease, the poor record of workers' compensation in dealing with occupational disease explains why few workers turn to this system for compensation. Because the expected value of recovery is higher under third-party tort actions, plaintiffs' attorneys have been encouraged to pierce the veil of workers' compensation by utilizing the tort system to sue manufacturers of hazardous materials under product liability suits.

The attributes of the tort system are almost a mirror image of first-party insurance. Wagner suggests that the tort system is ponderous, cumbersome, and uncertain. Its enormous delays are often due to the conscious manipulation of the attorneys. It incurs extremely high transactions costs. Indeed, the plaintiff recovers only about 40 percent of the outlays by the defendant, the remainder being absorbed by attorneys' and experts' fees. The tort system, however, provides the possibility of broader coverage.

## OPTIONS FACING SOCIETY

### Goals of Risk Management

The options facing Mary Peters are limited by the compensation regimes that currently exist. As a society, we face the question of whether these mechanisms are adequate and whether new options are desirable. To deal with these broader questions, we should consider how alternative regimes meet the objectives of societal risk management. In addition to just compensation, these objectives include risk spreading and efficient deterrence. The alternative compensation systems should be evaluated for their ability to achieve all of the objectives, subject to reasonable transactions costs (Table 10.1).

Risk spreading and compensation have been discussed in some detail, but the concept of efficient deterrence merits some elaboration. Efficient deterrence has at least three aspects. First, risk management is efficient if a given probability or severity of loss from some risky activity is achieved at a minimum cost. This

**Table 10.1**
**Societal Risk Management: Regimes and Objectives**

| OBJECTIVES | Statutory Regulation | Tort System | Administrative System | Insurance--First Party | Insurance--Third Party | Risk Charge |
|---|---|---|---|---|---|---|
| Risk spreading | No | Enterprise liability | Yes | Yes | Yes | Yes |
| Efficient deterrence |  |  |  |  |  |  |
| Cost effectiveness | Unlikely | No effect | No effect | No | Yes | ? |
| Minimum cost accident & prevention | Unlikely | Hand rule | Yes | No | Yes | ? |
| Balance risk/benefit | Sometimes prohibited | Strict liability | Yes | No | Yes | Yes |
| Just compensation | No | Subject to financial responsibility | Yes | No | Yes | ? |
| Transactions costs | HI | 60% | 30%+ | 40%+ | ? | ? |

aspect of efficiency is equivalent to cost-effectiveness. The market generally provides incentives for cost-effectiveness, although regulations may prescribe certain activities that are not cost-effective.

A second aspect of efficiency is balancing the expected marginal cost of accidents with the marginal cost of accident prevention. This rule minimizes the sum of the cost of accident plus accident prevention. In a tort system, this rule is embodied in the definition of negligence espoused by Judge Learned Hand. This rule is not sufficient to determine whether a risk is worth living with in the first place.

The third aspect of efficiency is a balance of risks and benefits. Even if a risky technology is managed on a cost-effective basis and a corporation balances the marginal cost of accidents and accident prevention, the costs of the activity may exceed the benefits. This is possible where a substantial portion of the costs are accidents borne by third parties.

## Risk-Management Regimes

Let us consider four ideal risk-management regimes, which are not necessarily mutually exclusive. A regulatory regime functions by imposing constraints on activities prior to an accident. Both tort and administrative regimes provide mechanisms for obtaining compensation after an accident occurs. A market-oriented insurance regime functions by imposing a charge on risk-bearing activities.

A regulatory regime spreads the risks of hazardous technologies by imposing burdens on industry. These costs are then passed on to the consumers of the industry's products. While regulatory regimes attempt to deter accidents by direct control, they pay little attention to efficiency. Regulations offer no incentives to minimize the costs of a given risk or to minimize the costs of accident versus accident prevention. Some laws governing hazards mandate a balancing of risks and benefits, but others prohibit consideration of benefits. Finally, the regulatory approach offers no compensation whatsoever.

The tort regime favors risk spreading through the theory of enterprise liability. According to this emerging doctrine, liabilities are shifted from victims to corporations because the latter are in a better position to spread risks to customers.

The tort regime offers no additional incentives for cost-effective risk management; however, neither does it impose constraints that could interfere with normal market incentives for minimizing the cost of achieving a given level of expected loss. The Hand rule of negligence provides incentive to compare costs of accidents versus prevention, but allows some losses to befall victims. In other words, if the cost of preventing a chemical accident exceeded the expected losses, a business would not be expected to prevent the accident nor compensate the victims. Such a rule provides imperfect internalization. In the case of hazardous chemicals, the courts have almost universally adopted a rule of strict liability, which provides much stronger incentives for companies to balance the marginal costs and benefits of risky activities.

The explicit purpose of the tort system is to provide just compensation. How well this system functions in the case of chemical injuries is debatable. While Wagner suggests that current tort mechanisms work poorly in compensating victims, Billauer et al. argue that the tort system works too well. According to the latter, the tort system all too frequently provides compensation for questionable claims. Not only does such overcompensation lack equity, it imposes a great cost on other social objectives. Overcompensation diminishes the ability of businesses to plan for foreseeable consequences of actions, businesses will lack the funds to compensate legitimate victims, and their products will be excessively priced. Not surprisingly, underwriters shy away from insuring the liabilities of hazardous chemical production. If so, then an additional harmful consequence may be that the public is deprived of needed products.

Without comprehensive data, it is impossible to compare the proportion of third-party injuries that have not been compensated to the proportion of compensated parties who were excessively compensated. As Strock suggests, the underlying reality may be less important in a policy sense than what the public communicates to their congressmen. Indeed, Congress has yet to resolve the conflicting strains from constituencies who perceive they have been undercompensated (such as victims of nuclear testing and Agent Orange) and those who perceive that victims have been overcompensated at their expense (like product manufacturers and insurers).

An administrative compensation system, like workers' compensation, is a variant of the tort regime with a lessened burden of proof and a narrower range of compensable damages. Administrative systems exclude compensation for general damages, like loss of consortium or pain and suffering. To the extent that appropriate employers are penalized for occupational diseases, the deterrent effect can be substantial. While in practice administrative schemes have not functioned well for such diseases, Locke argues that the workers' compensation system can be ameliorated to serve as the primary basis of compensation for occupational disease. He looks to simpler rules of liability and apportionment.

As both Katzman and Strock argue, there is little to suggest that the workers' compensation system provides a useful model for compensating victims of environmental disease. The black lung disease compensation system has had difficulty striking a balance between overcompensating and undercompensating victims. In dealing with a potential victim pool that can be circumscribed by exposure to coal mining, the Black Lung Fund at least has both theoritical limits to compensation and a defined list of potential parties against whom damages can be assessed. An environmental compensation fund would have no such limits since everyone lives in "the environment." Such a fund would become indistinguishable from first-party insurance, except that the "premiums" would be paid by the broader public, not the victim.

Basing societal risk-management strategy on first-party insurance is particularly inappropriate where the injury is due to circumstances firmly under the actual or potential control of another party. Such a regime would have no deterrent value. First-party insurance provides a mechanism for risk spreading for the same limited damages as administrative schemes. Nevertheless, in the guise of national health systems, first-party insurance has become the remedy of choice in most industrial democracies.

A third-party, or liability, insurance system is not a self-standing risk-management option, for it must be linked to either a tort or administrative system of compensation. Financial responsibility requirements which mandate liability insurance are essential in ensuring that a victorious plaintiff will indeed be able to collect. To the extent that premiums reflect expected losses, a liability insurance requirement can encourage efficient deterrence.

Finally, a quasi-insurance mechanism like a risk charge might be imposed on hazardous technologies, like chemical production or waste disposal. Indeed, Superfund at one time included both mechanisms, although they have been virtually eliminated as significant incentives. A risk charge could be earmarked for cleanup or victim compensation, but might also accrue to general revenues. The effectiveness of a risk charge as a deterrence would depend upon how closely its magnitude correlated with expected third-party damages. Improving this correlation would obviously increase the transactions costs of administering a risk-charge regime. As noted by Strock, the Japanese impose a risk charge, which is dedicated to a victim compensation fund.

The transactions costs of the various regimes are difficult to determine. These include the fixed public costs of system overhead (regulatory bureaucracies, courts), fixed costs of private risk management, and the variable costs of administering compensation if injuries occur. The overhead and fixed costs of alternative regimes have not been subject to serious quantification. In a regulatory regime, additional hidden costs include distortions caused by cost-ineffective constraints.

The variable transactions costs of the tort regime have been fairly well established by the RAND Institute for Civil Justice at approximately 60 percent of defendant's outlays. As noted by Locke in Chapter 7, variable transactions costs under workers' compensation are only slightly less. Of each premium dollar paid by employers for workers' compensation insurance in general, only sixty to eighty cents return to victims in terms of compensation, the remainder being absorbed by insurance administration or "loadings." In addition, a litigant's attorney may absorb as much as 15 percent of the recovery. If these percentages apply to occupational disease litigation, the total transactions costs may exceed 30 percent of the employer's outlays for workers' compensation." Loadings for first-party medical and property insurance are of a similar order of magnitude, reflecting the costs of risk analysis, premium processing, and claims settlement.

## Coping with Scientific Uncertainty

The major premise of this volume is that decisions about compensating victims of most latent chemical diseases are being made in a context of very weak scientific knowledge. With the exception of asbestos-related diseases, there may be uncertainty about (1) whether an individual is truly injured, and (2) whether there is a causal relationship between particular exposures to particular chemicals and particular injuries. As has been pointed out in Chapter 6, the system for compensating victims of asbestos exposure, where causality is now well established, provides little assurance that compensating victims of latent disease can be expeditious. For poorly understood chemical exposures, a compensation system is not likely to work even this well. How well do the alternative compensation regimes suit the conditions of scientific uncertainty?

Scientific knowledge is obviously essential in determining whether an injury exists. As Cothern and Schnare indicate, defining the existence of a disease is not straightforward. The scientific possibility of identifying subclinical injuries traceable to chemical damage may increase the range of compensable damages.

Once the fact of an injury is established, the scientific knowledge necessary to effect compensation is not necessarily the same as that necessary to effectively deter. Of the three compensation systems, first-party insurance has a decisive advantage in dealing with uncertainty about causation. Once the existence of an injury is established, first-party insurance compensation ensues without having to identify the causal agent or the pathways by which the victim was exposed.

Deterrence cannot be effective unless the generic causal links between the chemical exposures and the injury is known. Deterrent strategies do not require linking particular injuries to particular release or particular exposures. A chemical may be banned or its discharge taxed because of the suspicion of its deleterious health effects, without identifying pathways through which particular individuals are injured. Whether regulators base their decisions on "most likely" or "worst-case" dose-response relationships is a trans-scientific, political judgment. Conservative biases may result in overly cautious regulations, but at least they provide clear expectations for business decisions about what chemical products and what level of effluents are acceptable.

The tort and administrative compensation systems, which also have deterrent functions, must bear the greatest knowledge burden. In addition to establishing the existence of injury and the generic causal link between chemical exposure and injury, a specific exposure must be identified. The less certain the scientific knowledge, the higher the transactions costs in utilizing either of these compensation systems. Conservative biases in interpreting dose-response curves can destabilize expectations and hinder effective business decision making. Because tort and administrative systems are invoked after an injury is discovered, a business can never know what worst-case judgments about causality will be made in the future. The remote fear that its chemicals will some day be judged as causing injury can discourage the production of valuable products.

What little we know about alternative compensation systems suggests that we are not yet informed enough to identify an "optimal" compensation policy. Nevertheless, scientific uncertainty about chemical causality suggests the strategy of uncoupling compensation and deterrence regimes for chemical diseases.

# Index

Agent Orange, 6, 8, 68–69, 92, 146, 153
Allen v. United States, 93, 152, 154
Arnett v. Dow Chemical Co., 73, 76
Asbestos disease, 6, 69, 71, 73, 76–78, 85
Atmospheric Nuclear Testing Compensation Act, 154
Ayers v. Jackson Township, 6, 74–76, 130, 165

Black Lung Program, 8, 127, 140, 152, 157; Workers' compensation and, 109

Cancerphobia. See Health effects, mental distress
Causality. See Evidence of harm and health effects
CERCLA. See Comprehensive Environmental Response Compensation and Liability Act
CGL. See Insurance, CGL
Chemical exposure: assessment of, 44–46, 57; the chemical industry, 129; chemicals in humans, 3, 53–54, 59; chemicals in use, 3; people exposed, 2–3; relationship to injury, 3, 17. See also Health effects; Toxicology
Claim splitting. See Compensation
Clean Water Act, 126
Colucci, A. V., 13
Compensation: administrative, 155; bankruptcy and, 89, 136, 138; calculability, 130, 167; claim splitting, 77; congressional proposals, 149, 155–158; magnitude of, 5–6, 8, 111, 129, 130, 153, 156; medical surveillance, 73, 74, 157, 166; relationship to harm, 20, 90, 93–94, 109–110, 115–118, 132, 155, 157, 167; timing of loss, 131; typical scenarios, 2, 119, 154; uncertainty of, 5, 89, 130; for veterans, 153, 154; as victim assistance, 145–146; without injury, 74, 76, 77, 148. See also Workers' compensation
Comprehensive Environmental Response, Compensation, and Liability Act, 125, 126, 140, 145, 149; 301(e) study group, 149, 151, 153
Comprehensive General Liability. See Insurance, CGL

*County of Anderson, Tennessee v. U.S. Gypsum Co.*, 76, 77
Curry, Judge, 77

*Dartez v. Fibreboard Corporation*, 71–72
DDT, 3, 6, 54
Deep Water Port Act, 126
*Devlin v. Johns-Manville Corp.*, 69
Disease: defined, 14; hierarchy, 14–15; sentinels, 16. *See also* Health effects
Dubos, Rene, 54

*Eagle-Picher Industries, Inc. v. Cox*, 78–79
Economic analysis. *See* Risk management
EIL. *See* Insurance, EIL
Environmental impairment liability. *See* Insurance, EIL
Epidemiology. *See* Toxicology, Epidemiology
Evidence of harm: admissibility of, 67, 91, 152; burden of proof, 67, 90, 93, 94, 106; causality, 66, 70, 85, 90, 93–94, 109–110, 115–118; medical certainty, 90; reasonable probability of, 73, 75, 86, 87, 90; reliability of, 68, 70; rules of, 91, 148, 152; sufficiency of, 65, 67, 69, 70, 73, 92; "unreasonably dangerous," 93–95
Exposure: chemical, claims for, 5–6; extent of, 3, 5, 53; insurance concept, 125, 129; risk analysis assumptions, 46

Federal Employees Compensation Act, 152
Federal Insecticide, Fungicide and Rodenticide Act (FIFRA), 126
Financial Responsibility, 126–127, 129
*Friends for All Children v. Lockheed Aircraft*, 74
*Frye v. United States*, 68

*Gardner v. Hecla Mining Co.*, 70
*Gideon v. Johns-Manville Sales Corp.*, 71, 73, 77–78
Goldberg, L., 15
Goldman, Dexter, 13

Government: data needs, 8; role of, 8, 146

Hanson, H., 19
Health effects: adverse effect, 15, 43, 47; causation, 19, 41, 46, 66, 70; chronic versus acute, 6, 16, 42, 46–47; data on, 4; diagnosis (*see* Medical treatment); etiology (*see* Health effects, causation); hierarchy of, 14, 55; mental distress, 71; probability of harm, 20, 65, 67, 69–70; progression of, 16, 17, 54; similarity of, 16–17; symptoms, 17; timeframe, 15, 42, 75, 86, 88. *See also* Chemical exposure; Evidence of harm; Toxicology
*Herber v. Johns-Manville*, 69
Hormesis, 43

*In re Agent Orange Product Liability Litigation*, 92
Insurability, 129, 137–138
Insurance: availability of, 6, 126, 129, 134; catastrophic, 158; CGL (Comprehensive General Liability), 128, 131, 132, 134, 136, 137, 139; as a deterrent, 126–127, 128, 139; EIL (Environmental Impairment Liability), 131, 134–137, 139; innovations, 137; moral hazard, 132, 137; role of, 10, 126, 127; strains on, 133. *See also* Liability
Interactions, chemical, 41–42

*Johnson v. United States*, 90
Joint and several liability, 135–136

*Kehm v. Procter & Gamble Manufacturing Co.*, 68

*Lasswell v. Brown*, 75
Latency, 42
Legislation: need for, 145. *See specific laws*
Liability: assessment of, 89, 92, 106; congressional proposals, 150, 151; joint and several, 135–136; limitations of, 110, 132, 138, 150,151; negligence and, 93–94; predicting, 86, 91, 93

106–107, 136; proportional, 89, 118; theories of, 134–135; transfer of, 107, 135; who pays, 106. *See also* Insurance, compensation
*Lima v. United States*, 91

Medical treatment: biological basis of, 54–55, 59; efficacy of, 60–62; diagnosis, 19, 55, 57; medical specialities, 4–5; philosophy of, 56–58; surveillance, 73–74; treatment regimen, 57, 168
Medicine: allergists, 4; environmental, 5; occupational, 4
Mello, N. K., 15
*Mink v. University of Chicago*, 76
Moral hazard, 128, 132
*Morrissy v. Eli Lilly & Co.*, 76
*Mullaney v. Goldman*, 69

National Academy of Science (NAS), 18
National Toxicology Program, 18

Outer Continental Shelf Act, 126

*Parker v. Employers Mutual Liability Ins. Co. of Wisconsin*, 70, 89–90
PBB (polybrominated biphenyl): effects of, 15–16; human burdens of, 17
PCB (polychlorinated biphenyl), 54
*Plummer v. Abbott Laboratories*, 76
Price-Anderson Act, 145, 150
Professions, relationships between, 4, 7, 13, 147
Prosser, Dean, 74

RCRA. *See* Resource Conservation and Recovery Act
Reliability of evidence, 68
Resource Conservation and Recovery Act, 125, 126, 129
*Rheingold v. E. R. Squibbs & Sons, Inc.*, 76
Right-to-know, 158

Risk management: alternative approaches, 169, 170–173; benefit-cost analysis, 34–35, 94–95; limitations to, 35, 43; objectives of, 170; risk assessment, 66, 70, 74, 75; risk communications, 4, 49

Scharnweber, H. C., 16
Simon, W., 15
Spyker, J. M., 16
Sufficiency of evidence, 69
Superfund. *See* Comprehensive Environmental Response, Compensation and Liability Act

TCE (trichloroethylene), 35
Thresholds, 43
Toxicology, 4; biological model of, 16; data needs, 13, 18, 19; epidemiology, 41, 65, 66, 70, 74, 85, 90; defined, 66; as evidence, 70–71; role of, 9, 13; subliminal, 15, 56. *See also* Health effects
Trans-Alaska Pipeline Act, 126
Trans-science, 7, 86; definition of, 87

Uncertainty: contributions to, 32–33, 50; data limitations, 18, 45; effect on the court, 92; means to reduce, 41–49, 173; professional biases, 5, 8, 49; policy and, 31, 35, 50, 68

Veterans' Compensation and Pension Program, 152

Weinstein, Chief Judge, 68, 69, 92
Weiss, B., 15
Workers' compensation: applicability, 5, 70–71, 105, 109, 111–112; coverage of, 108–109, 110; history of, 107–108, 127; limitations of, 110, 112; magnitude of, 106; new proposals, 112–115; philosophy of, 105, 107, 110

# About the Contributors

BARBARA P. BILLAUER has established a highly successful legal practice which includes environmental and chemical victim law. Currently a partner at Stroock & Stroock & Levan in New York City, she is also an adjunct Associate Professor of Health and Safety at New York University Graduate School.

C. RICHARD COTHERN, Ph.D., is an Executive Secretary for U.S. EPA's prestigious Science Advisory Board. He has established himself through a series of highly regarded papers on risk assessment and risk management, with special emphasis on radioactive chemicals. He is the editor of a recent book on regulating radionuclides in drinking water.

KAREN L. I. GALLINARI is at the beginning of a legal career which will include occupational, industrial, and environmental health law. She is currently a member of Anderson Russel Kill & Olick in New York City.

MARTIN T. KATZMAN, Ph.D., has held various tenured faculty positions, most recently at the University of Texas–Dallas. His academic work has focused on the intersection of economics and environmental protection, especially with regard to the use of insurance. He is currently at Oak Ridge National Laboratories where again his research deals with environmental economics.

## About the Contributors

LAURENCE LOCKE, A.B., LL.B. Harvard Law School, 1942, Senior Member of Laurence Locke & Associates, Boston, Massachusetts, is the author of the Massachusetts Practice Series volume on Workmen's Compensation (West 2d ed. 1981) and has been a practitioner in the area of workers' compensation since 1946. He has served as legal counsel to the Massachusetts AFL-CIO on matters pertaining to workers' compensation and has authored a chapter on the topic in the *Annual Survey of Massachusetts Law*, published by Boston College Law School, for the past twenty years.

AVRAHAM C. MOSKOWITZ, formerly in private practice dealing with environmental and occupational law, is now Assistant United States Attorney for the Southern District of New York.

DAVID E. ROOT, M.D., MPH, formerly with the Air Force, has served on the NASA medical support team. He currently maintains an occupational and industrial medicine practice in Sacramento, California.

DAVID W. SCHNARE, Ph.D., MSPH, is Chief of the Economics and Policy Analysis Section within U.S. EPA's Office of Drinking Water. Author of a variety of papers on treating the chemically exposed, and on the development of policies regarding regulation of environmental chemicals, he earned his doctorate in environmental management and his master of science in public health in environmental science at the University of North Carolina at Chapel Hill. He was awarded a LEGIS Fellowship with the United States Senate, serving with the Appropriations Committee.

JAMES M. STROCK is the General Counsel for the U.S. Office of Personnel Management. At the time his contributions to this book were prepared, he was Special Counsel to the Committee on Environment & Public Works, U.S. Senate. He has also served as a Special Assistant to the Administrator of the U.S. EPA.

WENDY E. WAGNER is an attorney in the Environmental Enforcement Section, Division of Land and Natural Resources, for the U.S. Department of Justice. Previously a Law Clerk with Chief Judge Engel, U.S. Court of Appeals for the Sixth Circuit, she was on the Law Review at Yale Law School and earned an MES at the Yale School of Forestry and Environmental Studies.